Companion Planting for Beginners

A Comprehensive Guide to Growing Vegetables, Fruits, Flowers, Herbs, Cacti, and Succulents while Maximizing Yield and Plant Compatibility

© Copyright 2024 - All rights reserved.

The content contained within this book may not be reproduced, duplicated, or transmitted without direct written permission from the author or the publisher.

Under no circumstances will any blame or legal responsibility be held against the publisher or author for any damages, reparation, or monetary loss due to the information contained within this book, either directly or indirectly.

Legal Notice:

This book is copyright-protected. It is only for personal use. You cannot amend, distribute, sell, use, quote, or paraphrase any part of the content within this book without the consent of the author or publisher.

Disclaimer Notice:

Please note the information contained within this document is for educational and entertainment purposes only. All effort has been executed to present accurate, up-to-date, reliable, and complete information. No warranties of any kind are declared or implied. Readers acknowledge that the author is not engaging in the rendering of legal, financial, medical, or professional advice. The content within this book has been derived from various sources. Please consult a licensed professional before attempting any techniques outlined in this book.

By reading this document, the reader agrees that under no circumstances is the author responsible for any losses, direct or indirect, that are incurred as a result of the use of the information contained within this document, including, but not limited to, errors, omissions, or inaccuracies.

Table of Contents

PART 1: ORGANIC COMPANION PLANTING FOR BEGINNERS 1
  INTRODUCTION 2
  SECTION ONE: GETTING STARTED 3
  CHAPTER 1: THE BENEFITS OF ORGANIC COMPANION
  PLANTING 4
  CHAPTER 2: PLANNING YOUR GARDEN 10
  CHAPTER 3: TOOLS FOR ORGANIC COMPANION PLANTING 19
  SECTION TWO: PLANT SELECTION AND PAIRING 26
  CHAPTER 4: COMPANION PLANTING WITH VEGETABLES 27
  CHAPTER 5: COMPANION PLANTING WITH HERBS 48
  CHAPTER 6: COMPANION PLANTING WITH FLOWERS 63
  CHAPTER 7: COMPANION PLANTING FOR PEST CONTROL 68
  CHAPTER 8: SEEDS VS. STARTERS 78
  SECTION THREE: PLANTING, CARE AND MAINTENANCE 84
  CHAPTER 9: START WITH THE SOIL 85
  CHAPTER 10: PLANT THOSE PAIRS 95
  CHAPTER 11: WATERING AND CARING FOR YOUR PLANTS 105
  CHAPTER 12: TROUBLESHOOTING COMMON COMPANION
  PLANTING ISSUES 112
  CHAPTER 13: HARVESTING YOUR ORGANIC COMPANION
  PLANTING GARDEN 120
  BONUS: ORGANIC FERTILIZER RECIPES 129
  CONCLUSION 133

PART 2: CACTI AND SUCCULENTS 135
　INTRODUCTION 136
　CHAPTER 1: THE FASCINATING WORLD OF CACTI AND
　SUCCULENTS 138
　CHAPTER 2: CACTI AND SUCCULENT SELECTION: WHICH
　SHOULD YOU CHOOSE? 155
　CHAPTER 3: PLANTING AND ARRANGING YOUR CACTI AND
　SUCCULENTS 173
　CHAPTER 4: WATERING WISELY: THE CORRECT WAY TO
　WATER CACTI AND SUCCULENTS 184
　CHAPTER 5: CACTI AND SUCCULENT HEALTH, CARE, AND
　MAINTENANCE 194
　CHAPTER 6: PRUNING AND SHAPING STUNNING SUCCULENTS 204
　CHAPTER 7: PROPAGATION TECHNIQUES AND MAXIMIZING
　YIELD 215
　CHAPTER 8: PEST CONTROL, DISEASE MANAGEMENT, AND
　OTHER CHALLENGES 225
　CHAPTER 9: COMPANION PLANTING WITH CACTI AND
　SUCCULENTS 236
　APPENDIX: A-Z OF CACTI AND SUCCULENTS: SPECIES
　IDENTIFICATION REFERENCE 248
　CONCLUSION 254
HERE'S ANOTHER BOOK BY DION ROSSER THAT YOU MIGHT
LIKE 256
REFERENCES 257

Part 1: Organic Companion Planting for Beginners

An Essential Guide to Growing Vegetables, Fruit, Flowers, Herbs, and More for Maximum Yield and Quality

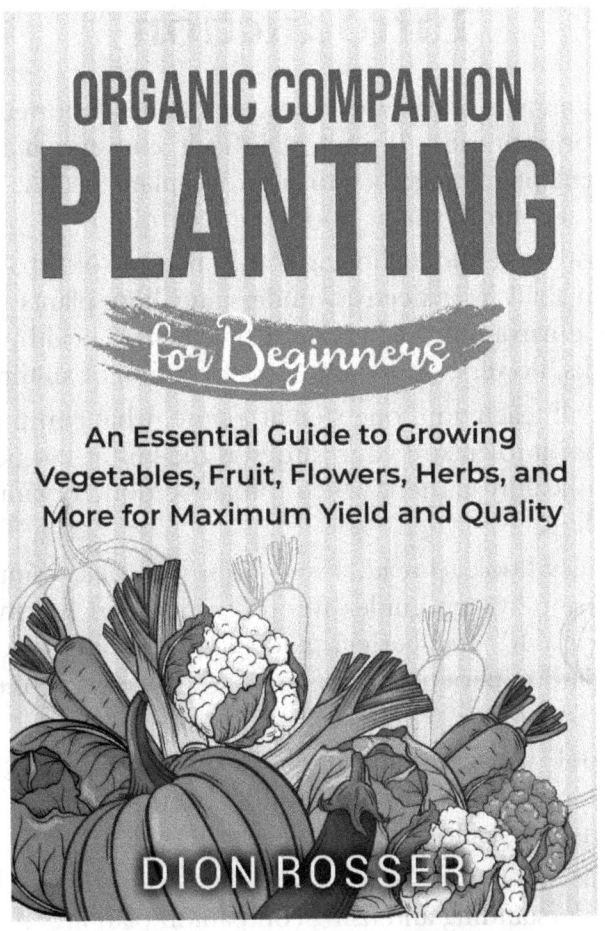

Introduction

Companion planting is one of the oldest techniques followed by gardeners and farmers for centuries. It isn't a difficult concept; it simply means planting different plants together to improve plant health, soil structure, productivity, pest control, shade, and weed control.

Gardening isn't just about shoving plants into any old space; there is far more to it than that! You need to understand how plants work together and with their environment to create a productive, healthy garden, be it vegetables, herbs, flowers, or, ideally, a combination of all three.

This book will teach you, one step at a time, what companion planting is and how to use it to ensure your garden is the best it can be. By the end, you will have a healthy garden, which will all be done organically, with no chemicals needed.

This is an easy book to read. I've written it in plain, simple language, with complete step-by-step guides where needed and full instructions on how to do things. It's the perfect guide for beginners who don't know where to start and experienced gardeners who need a refresher or more ideas.

This is a book you can buy once and keep forever, a guide you will refer to frequently – and you should. Even the most experienced gardeners still use books for information.

So don't wait any longer. Start reading and learn how to be a fantastic gardener by understanding and using companion planting.

SECTION ONE: GETTING STARTED

Chapter 1: The Benefits of Organic Companion Planting

To understand the benefits of companion planting, you first need to understand what it is. It isn't a difficult concept to grasp; it's simply planting different plants together for one or more benefits, such as health, growth, pest control, etc. Those are known as good companion plants, but some don't help each other and, in some cases, can even cause problems; these are bad companions.

A Brief History

Companion planting is a centuries-old technique dating back to when agriculture first began, and evidence has been found of it in ancient civilizations worldwide. Early farmers could see the benefits of growing certain plants together to get a bigger harvest, more fertile soil, keep pests away, and have a truly balanced, organic ecosystem.

Native Americans

The "Three Sisters" companion planting technique.
Anna Juchnowicz, CC BY-SA 4.0 <https://creativecommons.org/licenses/by-sa/4.0>, via Wikimedia Commons: https://commons.wikimedia.org/wiki/File:Three_Sisters_companion_planting_technique.jpg

Perhaps the best-known companion planting example originated with the Native Americans, who developed what is now a popular technique called "three sisters." This technique involves growing beans, corn, and squash together, each plant benefiting and supporting the others. The corn provides a trellis for the beans to grow, and the beans provide nitrogen for the soil, which helps nourish the squash and corn, while the squash acts as ground cover, stopping weeds from growing and keeping the soil moist. All three plants thrived, providing Native American communities with a nutritious and sustainable food source.

Ancient Egypt

Evidence of companion planting in ancient Egypt has also been found, where farmers grew plants like garlic and onions beside their barley. Because these crops gave off a pungent odor, they kept pests away and stopped the barley crop from being damaged or destroyed. They also used peas, beans, and other legumes as cover crops to make the soil more fertile for melons, cucumbers, and other climbing plants.

Ancient China

Companion planting was a key part of ancient Chinese agricultural practices. They grew many different plants together to provide support for climbers, control pests, and improve soil health. One example from ancient China was growing beans that fix nitrogen into the soil with grain crops, like millet and rice, helping to keep the soil fertile and increase yields.

Companion planting has evolved over the years as gardeners and farmers experiment with combinations to find new combinations, and it continues to evolve today. The more scientific knowledge we gain about how plants interact, the more understanding we gain of the many benefits companion planting brings.

In the 20th century, organic gardeners made companion planting popular, along with farmers who wanted an eco-friendly, sustainable approach to agricultural practices. Today, it is one of the most widely practiced forms of organic gardening, be it in the smallest backyards to the largest farms, enhancing the health, harvest, and strength of the crops while providing a chemical-free way of controlling weeds, pests, and fertilizing the soil.

The Benefits of Companion Planting

Companion planting offers plenty of benefits, the main ones being:

- **Disease Suppressant and Pest Repellent:** Some plants give off chemicals from their roots, flowers, or leaves, which keep pests away from nearby plants and suppress certain diseases.
- **Nitrogen Fixing:** Legumes, such as beans and peas, help fix nitrogen in the soil. Rhizobium bacteria is produced by the root system, which extracts the nitrogen from the air, embedding it in the earth to fertilize it. The bacteria give some of that fertilizer to the legume plant in exchange for sugars the plant produces by photosynthesis. This is called a symbiotic relationship because the bacteria and plant benefit, and the nitrogen in the soil also helps other plants nearby.
- **Trap Cropping:** These act as pest decoys. When one plant is more attractive to a certain pest or group of pests, it can be planted near plants that the pests attack. That way, the pests will

head for the trap crop and leave the main crop alone. Trap crops are nothing more than sacrifices; if they are perennial, they will come back the next year despite the damage caused by pests, or if they are annual, they'll produce seeds or seedlings. Some trap crops are known as "dead ends" because they kill the pest once they've been trapped.

- **Masking Scents:** Many animals and insects use smell to detect food. To stop pests from eating your flowers, plant other flowers with a stronger scent to mask them. These must be planted upwind of the main plant as the pests follow scent trails on the wind.
- **Camouflage:** Other pests use the physical shape of a plant to identify it as food. Plant pest-repellent companions among your crops to mask the shape of the target crop, and plant those that attract beneficial insects as further protection.
- **Stacking:** Another benefit of companion planting is creating protective environments that protect some plants from the cold, wind, or sun and support their growth. In permaculture, plants are layered, with tall ones at the back protecting shorter or lower-placed plants from the sun. That layer of plants then provides a sheltered area for groundcover plants – that way, every plant gets the ideal conditions to grow and thrive.
- **Nurse Cropping:** Similar to stacking, nurse cropping is about planting certain plants to protect smaller, more vulnerable ones from the strong sun as they develop. They also stop the soil from eroding and prevent weed growth.
- **Biodiversity:** Another important benefit of companion planting is biodiversity. By including a good mixture of plants in your garden, you create a strong ecosystem that can survive should a pest, disease, or bad weather weaken or kill off one variety. This provides security against the entire ecosystem collapsing when one type of plant fails.
- **Maximizing Space:** Instead of having huge gaps between plants, companion plants help you maximize your space – more plants, different species, all planted together.
- **Soil Health:** Some plants help fix soil health by producing certain nutrients. We mentioned beans and peas earlier – these add nitrogen to the soil, but other plants, such as radishes and carrots,

help keep the soil loose and free.

Attracting Beneficial Insects

While some plants are used to repel pests, another way of controlling unwanted insects is to attract the beneficial ones, along with birds and arthropods, including butterflies, spiders, centipedes, and beetles.

Beneficial insects help control certain pest species, and some of the best ones to attract include:

- **Pollinators**, such as bees and certain wasps
- **Predators** that feed on destructive pests – some of the more useful are hoverflies, lacewings, ladybirds, and praying mantis
- **Arthropods** that feed on pests – predatory mites and spiders are two you should be encouraging
- **Parasites** – these attack certain pests, such as some wasp species

However, if you want a garden filled with beneficial insects, you need to attract them, and that's where certain companion plants come in. There are two types of plants these insects need:

- **Nectary** – they provide nectar as a source of food
- **Insectary** – they provide permanent homes for beneficial insects to live in and spend the colder weather

Take a cornfield, for example. Nothing but corn, as far as the eye can see. It's a fantastic place for pests that feed off corn, but it does nothing to attract and support the beneficial fauna that feeds off those pests – there are no food sources and nowhere for them to live.

Plants with small, shallow flowers are ideal for beneficial insects – daisies, calendula, carrot, parsley, dill (when allowed to flower), and plants like sweet alyssum. And for these insects to find a permanent home, you must plant perennials. We'll talk more about these later.

Good and Bad Combinations

Many crops can be used as companion plants, but getting the right combinations is imperative because they all interact differently. Experience will be your guiding factor when deciding what works best for your garden, but here are some options to get you started:

Good:
- **Beans, Corn, and Squash:** These are the 'three sisters," so named because they work in unison. The corn grows tall, and the beans can wind around the stalks. The tall corn grows quickly and also provides shade for the squash. The squash grows close to the ground and stops weeds from attacking the other two, also adding much-needed nitrogen to the soil.
- **Aromatic Herbs and Cabbages:** Insects love cabbage, so plant strong-smelling plants and flowers to mask the scent of cabbage. Mint and rosemary are perfect, but any scented herb will work.
- **Sunflowers, Cucumbers, and Radishes:** Sunflowers, just like corn, provide shade to plants and flowers below, protecting them from the harsh sun. In return, radishes and cucumbers improve soil quality.
- **Tomatoes, Basil, and Marigolds:** Basil improves the taste of tomatoes, not only after growth but when paired together in soil. Marigolds attract bees to pollinate plants while repelling pests.

Bad:
Be careful because not all combinations of plants and flowers will work.
- **Tomatoes and Potatoes:** Both are closely related; when you plant the same family of plants too close to one another, they compete for nutrients and often attract the same pests, causing an overload.
- **Brassicas and Strawberries:** Brassicas include cabbages, cauliflower, and broccoli; strawberries stop them from growing well.
- **Beans and Onions:** These require very different conditions for growing, so planting them together may result in the onions slowing down the growth of the beans.
- **Cucumbers and Aromatic Herbs:** Some herbs can stunt the growth of cucumbers, and strong herbs can also change the delicate flavor of the cucumber fruit.

Keep reading, and you'll find all the information you need to be successful at companion planting.

Chapter 2: Planning Your Garden

Now you know what companion gardening is all about, it's time to start planning your garden. If you are new to this, you probably want to dive right in and get started, but there are things you need to do first.

You need to plan your garden, and for that, you need to choose a location. To do this correctly, there are a few factors to keep in mind:

1. Convenience

This is one of the most important factors when choosing a good location. If you don't have to walk too far or fight your way down to the garden, you'll have more success with growing. You'll notice watering requirements, pests, and other problems much quicker, and you'll get your harvest in on time, too.

2. Sunshine

Plants need sunlight.
https://unsplash.com/photos/dQejX2ucPBs?utm_source=unsplash&utm_medium=referral&utm_content=creditShareLink

Most plants need sunlight daily. Vegetables need at least eight hours – preferably more, to ensure they grow well and the fruit ripens. Monitor your garden to see how much sun it gets and where the partial and full shade areas are throughout the day.

Start by drawing out your garden plan and mark the sunset/sunrise hours. Head to the yard every hour and mark whether each is in full sun, shade, or partial shade. Count the hours each area is in the sun; any that don't get enough are unsuitable for vegetables.

3. Soil

Once you have decided where your garden is going, it's time to look at the soil. Healthy soil is critical for healthy plants. Ideally, you want well-drained, fertile soil. Learning about the soil in your garden is essential, so follow these steps:

Dig a Hole

Make it 12 inches by 12 inches and 6 inches deep. Put the soil you dig out onto a tarp or in a large bucket and look at it. Write down what you see:

- What colors are there? Is it loose soil or tight? Light or heavy?
- Rub a little between your fingers and write how the soil feels.

Count the Worms

Are there any worms in the soil you dug out? Have a good look – if there are at least 10, you have good fertile soil. If not, you'll need to learn how to improve the soil.

Test Drainage

Now, you want to see how well the land drains. Dig your hole 6 inches deeper – you need 12 inches for this test. Fill it with water and monitor how long it takes to drain away.

Once it has drained, repeat and monitor how long it takes to drain away again. If it's more than 8 hours, your soil drainage needs to be improved, or you could consider a raised bed or container gardening instead.

4. Water

You can use mulch and compost to help your plants become drought-resistant, but they will still need watering at some point, especially if you live in a zone that gets little rain and a lot of heat in the summer months. Seeds, in particular, need moist, warm soil for germination, and most vegetables need a steady water supply to ensure healthy growth. The ideal amount is an inch of water per plant per week.

Think about how you are going to water your garden. Is there a clean source nearby? Will you use a hose, watering cans, or drip hose? This must be considered before you get too far down the line of starting your garden.

5. Movement

The last thing to consider is how elements move through your yard. Some things to consider are:

- **Water:** How do snowmelt and rain flow through your land? Would too much wash your garden out? Does it run off or puddle and hang around, making things soggy?
- **Wind:** What direction does the wind blow in each season? Will it affect your garden, especially if it is a strong wind? Will you get weed seeds blown in from other gardens or fields nearby?
- **Equipment Access:** Can you easily access the area's necessary garden equipment? You'll need a wheelbarrow, possibly a tiller, and you might even need to have truckloads of compost unloaded and want easy access to it.

Once you have chosen your location, you should prepare it for planting.

6. Clear the Ground

Clear the ground of all grass and weeds completely, and remove any other debris and stones in the area. When planting in the fall, use layers of newspaper (up to 10) with layers of compost, potting soil, and topsoil. They can be layered or mixed. Water it well and leave it – by spring, you will have a weed-free area ready for planting.

7. Test/Improve the Soil

You can hire a professional to test your soil. However, you can invest in a kit to test your soil for smaller areas. This won't give you as much information, but it will give you a rough idea of whether your soil has sufficient nutrients or needs adjusting in some way.

Most of the time, soil in residential gardens needs a nutrient boost, especially if the topsoil has been removed for some reason. Low nutrient levels are only one thing; your soil may be poorly drained or compacted. Resolving this is easy enough; add lots of organic matter. Add a couple of inches of good compost when you till or dig a new vegetable bed. If you are working on an existing bed or don't plan on digging the soil, lay the compost over the top. Eventually, it will rot down and become organic material (humus). Earthworms will do your job for you and mix it in with the soil.

8. Prepare Your Beds

Before digging, decide what type of bed system you want – raised beds, straight rows, four-square, etc. Whatever system you go for, it's imperative to ensure the soil is loose – this will allow the plant roots to grow and pick up the nutrients and water they need. If you are planting straight into the ground, use a tiller to loosen it or get stuck in and dig it by hand. Tilling is ideal if you need to add ingredients/amendments to your soil, as the tiller will incorporate those. However, be aware that too much tilling can damage the soil structure. If your beds are small, stick to digging by hand.

You want to make this easy for yourself, so don't dig when the soil is too dry. It will be harder to get through. When the soil is too wet, it will be heavy, and it will only take more energy out of you. You want a little moisture in the dirt when you dig. Start with a garden fork to loosen the earth, and then dig it up with a spade. Turn the soil over and add in the organic matter. If you need to tread on mixed soil, lay down planks to distribute your weight.

Choose the Right Organic Fertilizer

Organic materials are great for the soil and are easy to source. Good fertilizers add nutrients gradually, working over a period of time to support plant growth. A decent product will feed your garden with macro and micronutrients, and you won't need to add chemicals.

Your plants need certain macronutrients which are found in most organic fertilizers, including the following:

- Calcium
- Magnesium
- Nitrogen
- Phosphorous
- Potassium
- Sulfur

These boost healthy growth and protect against some diseases that stunt development.

Your plants also need the following micronutrients, which are also found in organic fertilizers:

- Chlorine
- Copper
- Iron
- Manganese
- Nickel
- Zinc

These help the plants grow flowers, healthy leaves, and healthy green and yellow coloration.

This balance of macro and micronutrients cannot be found in chemical fertilizers, and chemicals don't stay in the ground long enough, so you become obliged to use them regularly – possibly causing more damage to the soil. Organic fertilizers are released slowly, working to improve water retention and soil quality over the long term.

They are also much cheaper, and you can even make your own from ingredients you already have in your home.

The Main Types of Organic Fertilizer:

Organic fertilizers can be produced from many sources, with the main ones being:

- Animal-based
- Mineral-based
- Plant-based

Animal-Based

These are typically made from animal manure and the remains left after slaughter, such as blood and bone. These are higher in nutrition than the other types and are best for leafy plants. Cow manure is the most common as it has a good balance of nutrients for all types of gardens and lawns.

Mineral-Based

These are produced from chemical processes using readily available elements from the environment. They are critical to rebalancing the composition of the soil by adding at least one macronutrient, depending on which fertilizer you use. Depending on how much you use, these can also help balance the pH level, but efficient use is required to do the most good without damaging the soil structure.

Plant-Based

As the name suggests, these are made from agricultural and plant by-products, such as molasses, green manure, cover crops, seaweed, cottonseed meal, and compost tea. They quickly break down, feed your garden plenty of nutrients, and help with soil regeneration and plant growth. They are the best choice if your garden soil is poorly drained.

How to Choose the Best

The best fertilizer is the one that matches your soil type, so you need to test your soil if you want to get it exactly right. A proper test will tell you:

- The macro and micronutrient levels in your soil
- What plants will thrive
- If you have balanced soil and, if not, what's needed to boost it.

The right organic fertilizer depends on your soil type, what you want to grow, and each plant's needs.

Garden Bed Layout Ideas

Much of how you plan your garden will come down to the available space. You will also plan around what you are planting and the maintenance needed. You can grow a garden that needs no tending, but it might not be what you want.

Rows

Rows are easy to tend to. They divide your garden neatly, and you generally run them from north to south, though you can also opt for east to west. As long as you have enough space between the rows, you can easily tend to your garden.

Tall plants like beans and corn should be planted at the north end so they don't shade other crops. Medium plants go in the middle, and smaller plants at the end. However, when you get into companion planting, this will change slightly.

Four Square

This is a simple layout, with a garden bed divided into four equal sections, each representing a separate bed. You don't need to mark these out physically if you don't want to. Each bed represents plants that require different amounts of nutrients.

Those that take a lot from the soil should be planted together and sparingly. Plants that take less can be planted together in numbers.

Rotate the crops after every season so the soil remains even over all planter boxes, and they each have the same nutrient needs. The layout is as follows:

HEAVY FEEDERS	MEDIUM FEEDERS
LIGHT FEEDERS	SOIL BUILDERS

After the first year's harvest, strip out the beds and prepare them for the next year. Each year, you will rotate the crops one square to the right, so in your second year, it will look like this:

LIGHT FEEDERS	HEAVY FEEDERS
SOIL BUILDERS	MEDIUM FEEDERS

And so on. This maintains balance in the beds.

Square Foot

Divide your grade into 4 x 4 sectors - the number of sections will depend on the size of your garden, and each will be one foot by one foot. Be sure to plant flowers that need support beside a wall or other structure. The key to this growing method is not to overcrowd each square, so be sure to check how many plants you can have in each sector.

Block

Block layouts are also known as close or wide-row planting, and they provide a much higher yield than standard row planting, with the added benefit of keeping the weeds down. The plots are similar to the square method, but the sectors can be as long as you like. This removes the need to add extra walkways, giving you more space for planting.

Using this method, you can plant a lot in a small space, but only if there is ample drainage and the plants are regularly watered. You need to tend to the plants regularly to ensure they grow and be careful to keep an eye out for pests. The rectangles can be up to 4 feet wide, but the length is only capped by your space. This makes them easy to weed and maintain. Keep walkways no more than 2 feet wide, and unless you make them out of paving slabs, add mulch to the walkways in the form of wood chips, grass clippings, or another type of organic mulch.

Make sure your plants are equally spaced in both directions. For example, a carrot patch would be spaced 3 x 3 inches. If you build a 3 x 3-foot bed, you can fit the equivalent of one 24-foot row of carrots into it – incredible space savings with a higher yield.

Vertical

Vertical gardens are ideal if you don't have much space. As the name suggests, you are planting upwards instead of horizontally. This can be done in vertical beds, baskets, or any other container that can hold soil vertically. A common method is to stack containers. This requires some work to set up but is easy to tend to once plants grow.

Containers or Raised Beds

These work well for smaller gardens or where your soil is too far gone to salvage. There are no limits to this kind of layout; the bonus of using containers is that you can move them around.

Chapter 3: Tools for Organic Companion Planting

Having the right tools is critical to getting your gardening done efficiently; it's all about making life easier so you have more time to enjoy the fruits of your labor.

Some things to consider when you purchase tools are:

- **Quality:** cheap tools won't last five minutes, so don't waste your time or money. Instead, purchase high-quality, well-made tools that will last longer. Tools often break at the joints (usually where the handle is attached), so look for one-piece tools that will last.
- **Materials:** When choosing wooden handles, hardwoods do not splinter like softwoods, so they will last longer and pose less of a hazard. Steel is great but heavy, aluminum is lighter but not as strong, and fiberglass is a good mix of the two.
- **Design:** consider ergonomic tools. For example, for those with cushioned or bent grips, consider the weight too- if they're too heavy, you won't be able to use them for too long, while tools that are too light are not likely to be strong enough.

Before shopping for tools, list everything you need and ensure you buy the best quality you can afford. Here are some ideas for useful tools:

Spade and Fork

A spade and fork are essential for marking beds and digging them.
https://unsplash.com/photos/vdD1rcsdL3E?utm_source=unsplash&utm_medium=referral&utm_content=creditShareLink

These are essential for marking your beds and digging them over. Spades can help you dig up hard soil and deep holes for trees and shrubs, while forks help you break the soil into a finer consistency. They come in all materials and sizes, so choose one that meets your requirements. Do not mistake a spade for a shovel; shovels have a flat-bottomed head, while spades are usually pointed and sharper.

Bucket

Good buckets are an excellent form of transportation. Carry your hand tools, mulch, compost, water, and even plants to where they need to go. If you can handle the weight, go for a galvanized aluminum bucket; if not, choose a sturdy plastic one.

Trug

This large, woven basket is ideal for holding your harvest and weeds, moving soil about, and much more. Some can even hold water.

Cultivator

A handle attached to a claw-like formation of metal prongs. You can use these to break up the soil, get large stones and rocks out of the ground, loosen plants for harvest, dig out weeds, and mix amendments into the soil.

You will find them in different materials, from plastic and wood to stainless steel and carbon fiber. If you want to go all out, you can opt for a two-sided tool with various tools on one side and a cultivator on the other. If you have a bad back or need to work on a larger area, you can purchase a tool with a long handle or grab an extender for your current cultivator.

Hand Fork

Excellent for loosening soil and digging over beds and containers to remove weeds, dig amendments in, and loosen the soil around plants to make for easier harvesting. Choose one made of strong plastic or metal to get the best results- some plastic ones are weak and won't last five minutes.

Footwear

The right footwear is important, so choose comfortable, durable, easy-to-clean shoes or boots. Try to have one pair dedicated purely to gardening.

Washable footwear is preferred when gardening.
https://unsplash.com/photos/tWE9W_5qTd0?utm_source=unsplash&utm_medium=referral&utm_content=creditShareLink

Non-washable shoes are not ideal as you can easily track pathogens from one garden area to another, risking plant disease. Plus, they'll get ruined fairly quickly. Clean off your footwear after every session in the garden to ensure they don't have any nasty diseases on them.

Garden Rake

Garden rakes have much firmer heads with short, strong tines, whereas a leaf rake is larger, lighter, and more flexible with long bent tines. A garden rake is dual-purpose. You can use the tined side to loosen weeds, roots, and rocks, remove dead grass, and spread the soil or soil amendments. The flat side can help you make furrows in the soil, smooth out the soil before planting, and lightly cover your seeds.

Rakes come in various sizes, so choose one that meets your requirements.

Gloves

Gloves are an important part of your gardening toolbox, but you can't wear just any old gloves – that means rubber kitchen gloves and woolly winter gloves are no good. Invest in a decent pair of proper gardening gloves. They are durable, breathable, washable, and provide a good grip. You may need a heavy-duty pair for digging, weeding, and heavy garden work and a lighter pair for sowing seeds and plants.

Wheelbarrow

A wheelbarrow can help you move equipment around the garden.
https://unsplash.com/photos/x6UXMqFw6GU?utm_source=unsplash&utm_medium=referral&utm_content=creditShareLink

Wheelbarrows make life easier because you can move equipment around the garden hassle-free. You can transport your tools, bags of compost, weeds, even buckets of water, and just about anything to help you with your gardening work.

Hoe

There are different types of hoes, and the one you choose will depend on your garden. If you focus on vegetables, you'll need a broad, strong hoe, while a perennial garden needs something lighter and thinner.

Hoes are great tools for removing weeds, especially between rows of plants, and prepping the soil in your garden beds. Traditionally, they have a flat blade with a sharp point to dig into the soil, although some have a flat bottom edge. You can also use a hoe to remove stones and rock, cover seed furrows, cut grass, make furrows, weed, and till the soil.

Hose

Lugging buckets down the garden will soon get tiresome, so invest in a good hosepipe with a multi-setting head. Make sure it is long enough to get where you need it to – you may need to join two or more together. The fittings are usually plastic, but invest in brass ones if you can get one – they last a lot longer. To save time, you can set up an irrigation system or run a soaker hose. Once these are set up, simply attach the hose and leave it to water your garden while you finish other tasks. These systems also use around 70% less water than standard hoses, and the water goes exactly where it is needed – the plant roots.

Moisture/Light/pH Meter

You can buy these separately or purchase one tool that does it all. All three are important for your garden. The pH meter tells you if the soil is right for the plants you want to grow, the light meter tells you if your plants are in too much sun or shade, and the moisture meter lets you know when it's time to water.

Tiller

These are standard tools for breaking the ground up, loosening soil, and making it easier for you to dig and plant. You can purchase manual, gas, or electric tillers, and what you buy depends on your garden. If your soil is hard and compacted and hasn't been touched in a long time, you will need a heavy-duty gas-powered tiller. However, if you have a small or medium-sized garden, you can use a smaller one to prepare the soil, remove weeds, and compost.

Pruners

Pruners are useful for cutting flowers.
https://www.pexels.com/photo/pruner-on-top-of-a-seedling-tray-6508421/

Another necessary item in your toolbox is a pair of pruners, otherwise known as *secateurs* or *shears*. These are useful for cutting flowers, pruning shrubs and plants back, and deadheading roses and other flowering plants (cutting off the old flower blooms to help signal the plant to grow new ones). Choose a good-quality pair with a sharp blade that produces a smooth, clean slice that helps the wound heal and keeps the plant healthy.

Tarpaulin

It's not necessarily a garden item, but tarps are good for many things. You can use them for covering materials and soil. Dragging plants to their new location (especially large shrubs), dragging rubble, leaves, and grass cuttings to where you want them, dragging soil or compost to the right place, storing soil you dug up while planting, lining your car trunk for when you bring plants home, and wrapping shrubs for the winter.

Trowel

An all-purpose tool, a trowel is a small spade used for small-scale cultivation. You can use a trowel for digging holes for planting, digging stones and rocks out, scooping compost into containers, doing small excavating, weeding, and transplanting. They come in all materials, different blade and handle lengths, and some have comfortable grips. Choose a full-tang trowel so it has less chance of breaking and bending.

You can also purchase trowels with measurement markings on the blade, ideal for helping you gauge how deep to dig a hole.

Other Tools

You should also consider having a companion planting chart as an easy guide to see what to plant with what. Garden planning software can help you plan out your garden, while gardening apps also help you plan and identify plants and weeds and give you plenty of advice on pests and beneficial insects. A soil test kit is useful if you don't want to send your soil off for testing, while a garden journal can help you keep track of what you have planted and where, dates, varieties, notes on germination, fruiting, pruning, etc., and notes on problems with pests and other garden issues.

Lastly, purchase a selection of spray bottles for your organic fertilizers.

SECTION TWO: PLANT SELECTION AND PAIRING

Chapter 4: Companion Planting with Vegetables

Vegetables love to grow with companion plants, benefitting from stronger plant growth, better flavor, more yield, and fewer pests and diseases. However, it's worth noting that, while companions work, each region will differ, as will each garden, so experimentation and knowledge are key.

This chapter will list the most popular vegetables, their favorite companions, and those you should avoid.

Asparagus

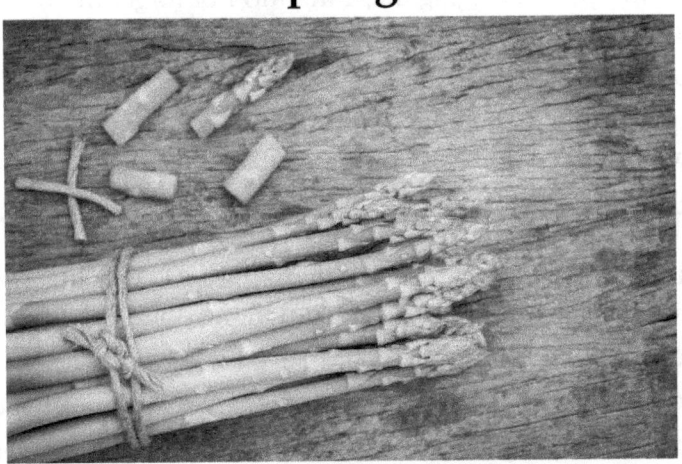

Asparagus takes a few years to be established.
https://www.pexels.com/photo/flat-lay-photography-of-asparagus-351679/

Ideal for patient gardeners, asparagus takes a few years to get established, but it's well worth the wait. An established, well-cared-for asparagus bed will reap rewards in pounds of delicious vegetables.

Good Companions:
- **Basil and Parsley–** encourages vigorous growth
- **Tomatoes –** deter asparagus beetles

The tomatoes also benefit from the parsley and basil with stronger growth and more flavorful tomato fruits. Basil also deters the tomato hornworm.

Asparagus will also grow well with marigolds, comfrey, dill, and coriander, as they keep spider mites, aphids, and other pests away – do your research, though, if you grow tomatoes with asparagus. The companions must be kind to tomatoes, too.

Bad Companions:
- **Garlic and Onions –** stunt asparagus growth
- **Potatoes –** deep-rooted like asparagus and compete for space and nutrients

Beans

A good crop to grow as they add nitrogen to the soil. Some gardeners harvest half their crop and then dig the rest in to add more nutrients to the soil, but you can achieve this by harvesting all the beans, pulling the plants, chopping them up, and digging them in – do not forget the roots, as this is where the nitrogen is stored.

Good Companions:
- **Squash and Corn:** the three form the 'three sisters.' Corn will grow tall, providing shade for squash and beans and a stem for the beans to grow around. Squash keeps weeds at bay, and beans add nitrogen back into the soil.
- **Marigolds:** perfect for reeling pests. African and French marigolds exude a chemical from their roots to deter nematodes.
- **Potatoes and Catnip:** deters Mexican flea beetle.
- **Rosemary and Nasturtium:** also deter Mexican flea beetles.
- **Summer savory:** repels flea beetles and induces strong growth and better flavor.
- **Eggplant, Radishes, and Cucumbers:** encourage growth.

Other good companions are celery, cauliflower, cabbage, broccoli, carrots, strawberries, and peas, and these also benefit from the nitrogen fixed into the soil by the beans.

Bad Companions:
- **Onion family:** this includes garlic, onions, leeks, and scallions, all inhibiting growth.
- **Kohlrabi, Basil, and Fennel** – also inhibit growth.
- **Sunflowers:** the toxins from the flowers inhibit bean growth.

Beetroot

Beets are incredibly easy to grow but require rich, fertile, well-drained soil.

Good Companions:
- **Brassicas:** the family that includes dense leafy vegetables like cabbage, broccoli, Brussels sprouts, and more. The beets provide minerals for the soil, which benefits the brassicas, and their leaves are also high in magnesium, making great compost for the brassicas.
- **Garlic:** a good deterrent against beetles, maggots, and moths. Garlic also contains an anti-fungal agent called sulfur, which protects the beets against fungal diseases. Just as it does in cooking, garlic boosts the flavor of beetroot when growing.
- **Mint:** improves beet growth, attracts predators to keep aphids away, and repels some rodents, fleas, and flea beetles. However, mint should be grown in containers as it is terribly invasive in the ground.

Bad Companions:
- **Pole Beans, Chard, and Field Mustard:** all of these will stunt the beet's growth

Broccoli

Broccoli requires regular feedings.
https://unsplash.com/photos/l55IGtwI8mI?utm_source=unsplash&utm_medium=referral&utm_content=creditShareLink

Broccoli is part of the brassica family, and it likes a lot of nutrients, which means regular feedings, especially with calcium.

Good Companions:
- **Aromatic Herbs:** rosemary, dill, potted mint, thyme, and basil all act as pest repellents.
- **Garlic:** also keeps pests away.

Bad Companions:
- **Other Brassicas:** all brassicas are heavy feeders and compete with one another for the nutrients in the soil, leaving the soil in very poor condition.
- **Asparagus, Pumpkins, Melon, and Corn:** as above, these are also heavy feeders.
- **Nightshades:** includes eggplant, peppers, and tomatoes; these will stunt its growth.
- **Strawberries and Pole Beans:** also stunt growth and compete for nutrients.

Because you can't plant much with broccoli, you tend to get too much wasted space. Maximize this by planting light feeders that won't compete for nutrients, such as marigolds, nasturtiums, bush beans, lettuce, shallots, and cucumber.

Brussels Sprouts

Also part of the brassica family, these are susceptible to many pests, making them frustrating to grow. They attract all sorts, from aphids and caterpillars to whiteflies – and many more.

Good Companions:

- **Nasturtiums:** repel some aphids, squash bugs, and whiteflies.
- **Basil:** repels mosquitos and flies and attracts beneficial insects like bees.
- **Garlic:** repels aphids and Japanese beetles and protects against blight; best planted in between Brussels sprouts for the best protection.
- **Marigolds:** repel plenty of pests.
- **Mustard:** a popular trap crop, mustard attracts many pests for Brussels sprouts. However, once attacked, the plant must be destroyed and replaced.

Bad Companions:

- **Pole Beans, Strawberries, and Tomatoes:** like all brassicas, these will stunt its growth.

Cabbage

Another brassica, cabbage is a beacon to many pests and, like all members of this family, it should be grown under fine mesh to reduce the attacks and stop butterflies from laying eggs on the leaves.

Good Companions:

- **Rosemary and Sage:** their scent repels the cabbage moth – plant these between the cabbage rows to repel pests and suppress weeds.
- **Chamomile:** improves flavor.
- **Marigolds:** repels cabbage moths, aphids, and many other pests.
- **Onions, Beets, and Celery:** repel pests and improve the flavor.

Bad Companions:
- **Tomatoes, Mustard, Grapes, and Pole Beans:** these will stunt the cabbage's growth.

Carrots

Carrots are relatively simple to grow, requiring little care except for watering and weeding.

Good Companions:
- **Tomatoes:** carrots break the soil up and aerate it, improving tomato plant growth. However, they must be planted at least 15 inches apart; otherwise, the tomatoes will stunt the carrot's growth. Tomatoes also shade the carrots and secrete solanine, a chemical that repels pests.
- **Onions, Leeks, and Garlic:** interplanting with these can help repel the dreaded carrot fly.
- **Aromatic Herbs:** chives, parsley, sage, rosemary, etc., will all repel pests.

Bad Companions:
- **Coriander and Dill:** both plants secrete chemicals into the ground that kill carrots.
- **Parsnips:** they also attract the carrot fly, so it's not a good idea to plant them together.

Plant beans in the bed the year before you plant carrots. They will fix nitrogen into the soil for the carrots to feed on. However, harvest the beans and remove the plants before planting the carrots. Otherwise, they produce too much shade and will crowd the carrots out.

Cauliflower

Yet another brassica family member that attracts a lot of pests. Rather than planting it in rows, cauliflower should be interplanted among other crops to disguise it and keep the cabbage moth away.

Good Companions:
- **Cosmos:** repels aphids and cabbage worms.
- **Nasturtiums:** good trap crop to attract aphids away from the cauliflower.

- **Fennel:** attracts parasitic wasps; these lay eggs beneath the cabbage worm's skin, thus killing them.
- **Celery and Oregano:** both repel many pests that attack brassicas.

Bad Companions:
- **Tomatoes:** because both are heavy feeders, they compete for nutrients, and neither will grow very well.
- **Strawberries:** as mentioned above, they compete for nutrients and stunt growth.
- **Other brassicas:** all heavy feeders and all attract the same pests.

Plant beans the year before to fix nitrogen into the soil for the cauliflower to feed on.

Celery

Celery is not the easiest to grow and requires a lot of water. This is why you shouldn't plant too many.

Good Companions:
- **Brassicas:** celery deters cabbage moth.
- **Leeks and Onions:** attracts insects that would attack the celery.
- **Cosmos, Snapdragons, Marigolds, Nasturtiums, and Daisies:** all repel pests and attract predators like parasitic wasps and other beneficial insects.
- **Peas and Bush Beans:** add nitrogen to the soil.
- **Tomatoes and Spinach:** provide shade to keep the soil moist.

Bad Companions:
- **Asters and Corn:** both attract diseases and harmful pests.
- **Potato and Parsnip:** heavy feeders that strip the soil of nutrients and encourage harmful pests.

Corn

Sweet corn is easy to grow but should be planted in blocks of four (not rows) because this encourages pollination.

Good Companions:
- **Beans and Peas:** they fix the nitrogen the corn needs.

- **Squash:** good ground cover to keep the moisture in and the weeds down.
- **Cucumbers:** interplanting cucumbers and corn deter raccoons.
- **Clover:** acts as mulch and a nitrogen fixer; however, be aware that clover can spread fast and will need to be controlled.

Bad Companions:
- **Tomatoes:** attract harmful pests, such as the corn earworm, which will destroy the corn crop.

Potatoes are in both camps. On one hand, the corn shades the potatoes, keeping the ground cooler and moist. However, both are heavy feeders- they will soon strip the soil and suffer unless fed regularly. Potatoes also attract a lot of pests that will eat the corn, including cutworms, potato aphids, and more.

Cucumber

Cucumber grows well in a greenhouse or poly-tunnel but is thirsty, so it won't benefit from being planted near other thirsty plants.

Good Companions:
- **Corn:** keeps raccoons away from the corn, and the cucumber will use the corn as a trellis.
- **Nasturtiums and Marigolds:** both repel harmful pests, including beetles and thrips.
- **Oregano:** repels insects.
- **Dill:** improves cucumber flavor.
- **Lettuce, Onions, and Radishes:** all repel certain insects and help improve growth and flavor.
- **Beans and Peas:** for their nitrogen-fixing abilities, especially when planted the year before.

Bad Companions:
- **Potatoes:** compete with cucumber for nutrients and water.
- **Sage:** stunts growth.
- **Tomatoes:** stunts growth and attracts harmful pests.

Eggplant

An eggplant can be grown in warmer zones.
https://unsplash.com/photos/8cqlBGw84oU

Known as *aubergine* in Europe, the eggplant is popular and has a long growing and fruiting season. It loves the sun, so it can easily be grown outside in warmer zones. In cooler climates, it must be grown in a greenhouse or poly-tunnel.

Good Companions:
- **Catnip:** deters flea beetles.
- **Hot Peppers:** secretes a chemical that prevents Fusarium diseases and root rot.
- **Sweet Peppers:** secretes fewer chemicals but has the same effect.
- **Pole Beans:** for their nitrogen, but don't let them shade the eggplants.
- **Bush Beans:** repel Colorado beetles.
- **Mexican Marigold:** also repels Colorado beetle, but it doesn't get on with beans; you'll need to plant these OR the beans, not both.
- **Thyme and French Tarragon:** repel harmful pests and garden moths.

- **Tomatoes:** similar growing requirements – don't plant too close together, though, or they will crowd each other out.

Bad Companions:
- **Geraniums:** hosts diseases that can affect the eggplant, like root rot and leaf blight.
- **Corn and Zucchini:** are both heavy feeders and will compete with the eggplant for nutrients.

Kohlrabi

Kohlrabi can be grown in cooler climates.
https://unsplash.com/photos/LYefL2BqtBY

This is a cooler weather crop and is part of the brassica family, which means it attracts a lot of different pests.

Good Companions:
- **Onions:** deters pests, including cabbage moths.
- **Lettuce:** deters earth flies.

Bad Companions:
- **Strawberries and Tomatoes:** both stunt the Kohlrabi's growth.

Leek

Part of the allium family, leeks are easy to grow and can be left in the ground until needed. You can grow leeks, garlic, and onions together, but monoculture growing can attract pests and diseases.

Good Companions:
- **Strawberries:** the strong smell from the leeks keeps pests away from the strawberries.
- **Apple Trees:** leeks also deter pests from the trees, and the apple trees improve the growth of the leeks.
- **Carrots:** this is a two-way relationship: the leeks deter carrot flies, and the carrots deter onion flies. Both crops also break the soil up, promoting good growth.
- **Parsnips:** the leeks deter pests from the parsnips.
- **Nasturtiums, Poppies, and Marigolds:** all repel pests.
- **Pepper and Tomatoes:** the leeks keep pests away from these plants and also help maximize space as they can be planted around peppers and tomatoes.
- **Beets:** similar care requirements and the leeks repel pests from the beets.
- **Celery:** both plants can grow together in trenches and have similar nutrient requirements. Leeks also keep pests away from the celery.
- **Brassicas:** do not compete for nutrients and water, and the strong smell from the leeks deters pests that attack brassicas.
- **Aromatic Herbs:** attract pollinators and deter some pests.

Bad Companions:
- **Beans and Peas:** stunt the leek's growth.
- **Asparagus:** care requirements are vastly different.

Lettuce

Easy to grow, lettuce is a popular crop among gardeners worldwide.

Good Companions:

- **Mint:** grown in pots to stop it from spreading, mint will repel slugs.
- **Onions, Carrots, and Leeks:** are all slower growing and struggle to compete with weeds; planting lettuce around these crops will smother weeds.
- **Radishes:** radishes make the lettuce taste better.
- **Cucumber:** improves flavor and provides shade for the lettuce, but don't let the cucumbers crowd the lettuce out; plant with radishes, too, as they deter cucumber beetle.
- **Strawberries:** improve the soil and bring in the beneficial insects and predators.
- **Basil:** improves growth and flavor.

Bad Companions:

- **Brassicas:** all heavy feeders and compete with lettuce for nutrients, stunting its growth.
- **Fennel:** stunts growth.
- **Parsley:** makes lettuce bolt (go to seed) very quickly.
- **Celery:** attracts the same diseases and pests as lettuce, causing damage to both crops.

Onions

Another easy plant to grow, you can start these from seed or buy seedlings.

Good Companions:

- **Brassicas:** onions repel cabbage maggots, cabbage loopers, and cabbage worms.
- **Carrots:** they help each other by keeping the onion and carrot flies away.
- **Lettuce, Strawberries, Tomatoes, and Peppers:** onions keep pests away from these crops, and they do not compete with the onions for nutrients.

- **Parsley and Mint:** repel onion flies; do grow mint in a pot, though, as it is incredibly invasive.
- **Chamomile:** attracts beneficial insects like pollinators and repels other pests; also improves onion flavor.
- **Cucumbers, Peppers, and tomatoes:** do not compete for nutrients, and the onions keep pests away from them.

Bad Companions:
- **Beans, Peas, and Asparagus:** all require different conditions to thrive, so planting with onions will not benefit any of them.
- **Other alliums:** attract the same pests, causing an infestation.

Peas

Another popular crop – peas are good companions for lots of crops.

Good Companions:
- **Corn:** the peas can use it as a trellis.
- **Green Beans and Carrots:** require similar conditions and have no adverse effects on each other.
- **Turnips:** peas feed the soil with nitrogen for the turnips to feed on, while turnips repel pests.
- **Basil:** repels pests, especially thrips, which can decimate your peas.
- **Lettuce and Spinach:** benefit from the shade thrown by the peas and the nitrogen.
- **Cauliflower:** also benefits from nitrogen.
- **Nasturtiums:** good trap crop to keep pests away from the peas.

Bad Companions:
- **Alliums:** stunt the growth of the pea plants.

Potatoes

Potatoes are a wonderful crop to grow and, if looked after properly, can provide you with plenty of new and maincrop potatoes to keep you going through the winter.

Good Companions:
- **Chives:** attract beneficial insects and predators to attack potato pests and improve growth.

- **Cilantro:** also attracts beneficial insects and predators, like ladybirds, which feed on Colorado beetle eggs, parasitic wasps, and hoverflies.
- **Horseradish:** produces odors and chemicals that improve disease resistance.
- **Parsley and Thyme:** improve the flavor and attract beneficial insects.
- **Mint:** attracts beneficial insects and predators.

Bad Companions:
- **Nightshade family:** this includes peppers and tomatoes, which are in the same family as potatoes and compete for water and nutrients and attract the same diseases and pests.
- **Cucumbers:** they make the potatoes vulnerable to blight and compete for nutrients.
- **Sunflowers:** exude chemicals that stunt growth and seed germination.

Pumpkin

Pumpkins are easy to grow.
https://unsplash.com/photos/T9pdHqCsyoQ

Pumpkins are a favorite with most gardeners, as they are easy to grow. However, they do spread quite rapidly, so ideally, they shouldn't be planted near many plants.

Good Companions:
- **Beans and Corn:** this is the *three sisters* method mentioned earlier. The squash provides the ground cover for the other two crops, keeping moisture in the soil and crushing weeds. Pumpkins should be planted last in this method when the corn is at least 24 inches tall. Otherwise, the pumpkin will affect its growth.

Bad Companions:
- **Potatoes:** Pumpkins can cause blight in the potato crop.

Spinach

Another simple crop to grow. Simply harvest the leaves as needed, and the plant will continue growing.

Good Companions:
- **Beans or Peas:** these provide shade for the spinach and add nitrogen to the soil.
- **Tomatoes and Cucumbers:** offer shade and don't compete for nutrients with the spinach.
- **Lettuce and Strawberries:** both of these boost healthy growth in the spinach.
- **Mint:** this deters snails and slugs, the biggest pest to spinach.
- **Onion:** repels pests.
- **Carrots:** help make the soil structure better.
- **Radishes:** repels flea beetles and aphids.
- **Cilantro and Dill:** attract beneficial predators to prevent pest infestations.

Bad Companions:
- **Potatoes:** a heavy feeder, the potatoes strip the soil of nutrients and water. They also attract insects that will feast on the spinach.
- **Fennel:** stunts growth.

Squashes

The squash family includes the zucchini – these are just as easy to grow as the pumpkins.

Good Companions:
- **Corn and Beans:** most squashes create amazing ground cover and repel weeds, boosting corn and bean growth.
- **Nasturtiums:** a trap crop that attracts whiteflies, aphids, flea beetles, and other pests that might attack the squash. Plant them a distance away from the squash- too close, and the pests will hop from the flower to the squash. These flowers also make the squash fruit taste better.
- **Radishes:** deters the squash vine borer.
- **Sunflower:** provides shade.
- **Marigolds:** attract beneficial predators and deter nematodes in the soil.
- **Borage:** repels the pests, and the leaves can be mulched back into the soil to provide calcium.
- **Aromatic Herbs:** mint, dill, parsley, oregano, lemon balm, etc. These all repel plenty of pests. Make sure you grow the lemon balm and mint in pots, or they will take over the garden.

Bad Companions:
- **Melons and Pumpkins:** these compete for nutrients and water and will attract diseases and pests.
- **Beets:** this fast-growing crop can upset the squash's root system and prevent it from growing properly.
- **Fennel:** stunts the growth.
- **Potatoes:** will steal all the nutrients in the soil.

Strawberries

Easy to grow, strawberries are a favorite and produce fruit throughout the season, depending on which varieties you have. The two main types are June-bearers, which produce an earlier crop but have a short fruiting season, and ever-bearers, which produce fruit throughout a much longer season. If you have a strawberry patch, you should place netting over it to keep the birds away.

Good Companions:
- **Alliums:** onions, chives, and leeks help repel pests and help keep diseases away. Let the chives flower, and they will attract beneficial pollinators like bees.
- **Asparagus:** they share the same growing needs, but their root structures differ, so they don't interfere with one another.
- **Spinach:** both have the same growing needs and are small enough to grow in the same bed.
- **Beans and Peas:** improve nitrogen in the soil and boost growth.
- **Yarrow, Dill, Borage, Catnip, and Thyme:** all attract beneficial pollinators and predators and repel other pests while boosting plant growth and crop yield.
- **Marigolds:** repels a lot of different pests. Stick to dwarf varieties; otherwise, they will crowd out your strawberries and produce too much shade.
- **Blueberries and Cranberries:** all like the same kind of soil, and the strawberries are a kind of mulch for the other fruit bushes.

Bad Companions:
- **Mint, Okra, Tomatoes, Cucumbers, Peppers, Potatoes, and Eggplant:** all of these are prone to a disease called verticillium wilt, which can destroy your strawberries.
- **Melons and Winter Squash:** also prone to wilt, and the vines will strangle your strawberry plants.
- **Cruciferous Vegetables:** this includes cabbage, broccoli, cauliflower, chard, collard greens, and Brussels sprouts, and they can all stunt the strawberry plant's growth. Plus, they attract unwelcome pests that can decimate your strawberry harvest.

Tomatoes

Another very popular crop. Although these have been included in the vegetable section, tomatoes are, strictly speaking, fruit.

Good Companions:
- **Basil:** improves plant growth and health, makes the fruit taste better, and repels many pests, including spider mites, hornworms, aphids, and whiteflies.

- **Borage:** improves fruit flavor and healthy plant growth while repelling the cabbage worm and hornworm.
- **Chives:** deter aphids and bring beneficial pollinators in.
- **Garlic:** deters spider mites – some people place garlic bulbs in the soil around their tomatoes to keep insects away.
- **French Marigolds:** deter slugs, nematodes, hornworms, and other nuisance pests.
- **Mint:** repels rodents, flea beetles, white cabbage moths, ants, aphids, fleas, and other pests.
- **Nasturtiums:** deter fungal infections and pests, such as aphids, squash bugs, beetles, and whiteflies.
- **Parsley:** attracts hoverflies, which feed on aphids and other pests.
- **Asparagus:** these work together; the asparagus keeps nematodes away while the tomatoes repel asparagus beetles.
- **Carrots:** break the soil up.
- **Roses:** tomatoes protect the roses from blackspot.
- **Gooseberries:** tomatoes repel insects that would attack the gooseberry bushes.

Bad Companions:
- **Brassicas:** all of these attract numerous pests that will attack the tomatoes and also stunt the growth of the tomato plants.
- **Corn:** attracts corn earworms and tomato fruit worms that will also attack the tomatoes.
- **Fennel:** stunts growth.
- **Potatoes:** tomatoes and potatoes can be affected by blight; if one gets it, the other will, too.

One other plant that can be both good and bad is dill. While it is a young plant, dill improves healthy growth in tomato plants, but it will stunt the growth when it gets older. If you want to grow dill with your tomato plants, ensure it is fully harvested while young.

Turnips

Turnips are part of the mustard family.
https://unsplash.com/photos/9c1f8Nae6j4

Also called rutabaga, turnips are a wonderful crop to grow. They are part of the mustard family and are biennial, which means they take two years to mature. The first year is spent growing the roots, leaves, and stems, while the second year produces the flowers and seeds.

Good Companions:

- **Brassicas:** the turnips are a trap crop in this case, attracting pests away from the brassicas.
- **Garlic:** turnip roots repel the borers that attack garlic, while the garlic repays this by deterring aphids, beetles, and onion flies from the turnips.
- **Beans and Peas:** add nitrogen to the soil, and because turnips are root crops and peas grow straight up, this companionship helps maximize space.
- **Nasturtiums:** attract pests away from the turnip and also attract beneficial predators and pollinators.
- **Mint and Catnip:** deters aphids and flea beetles and attracts beneficial predators and earthworms. Grow in pots as they are invasive, chop the mint leaves off regularly, and mulch them into the soil around the turnips.

- **Thyme:** deters cabbage whiteflies and attracts beneficial pollinators and predators.

Bad Companions:
- **Potatoes:** both root vegetables; these will compete for nutrients, water, and space and will hold each other back from growing.
- **Onions:** onions are generally great companion plants, but they aren't the best to be paired with turnips because the onion bulbs grow beneath the ground, and there is a space issue. However, plant them a few feet away, and you'll benefit from them repelling pests from the turnips.

Zucchini

Also called courgettes, zucchini are incredibly easy to grow, and provided you give them the right care, you'll be rewarded with a bumper crop of these powerhouse vegetables.

Good Companions:
- **Radishes:** deter squash bugs, cucumber beetles, aphids, and many other pests. Because radishes are a fast-growing crop, you'll need to do several plantings throughout the season to reap the benefits.
- **Garlic:** repels aphids.
- **Beans and Peas:** add nitrogen to the soil.
- **Marigolds:** deters many pests and attract pollinators.
- **Nasturtiums:** the ever-popular trap crop, these will sacrifice themselves to the many predators that attack zucchini plants. Their flowers will also attract beneficial pollinators.
- **Aromatic Herbs:** including lemon balm, mint, borage, oregano, parsley, catnip, and dill; these all deter pests and attract predators and pollinators.

Bad Companions:
- **Potatoes:** stunt growth and attract pests that attack the zucchini, notably the Colorado beetle.
- **Fennel:** stunts growth.
- **Melons:** take up too much space and will crowd the zucchini out. Both plants will also be competing for nutrition.

- **Pumpkins:** compete for nutrients, and because they are the same family, there is a risk of cross-pollination, resulting in a large crop but small fruits.

As you can see, the same names crop up repeatedly as good and bad companion plants for vegetables. Most strong-smelling herbs make excellent companions because they keep the pests away, and the humble nasturtium is an excellent trap crop, constantly sacrificing itself by attracting the pests away from the main plant. I must reiterate that if you use mint, lemon balm, or catnip, you must plant them in pots or risk them taking over your gardening and smothering everything else.

In the next chapter, let's look at herbs as companions in more detail.

Chapter 5: Companion Planting with Herbs

Herbs are popular plants to grow. They make great companions for other plants, and they are also simple to grow, need very little care, and can even be used in the kitchen – fresh, dried, or frozen. Most gardeners have herbs growing in their gardens, but the fact that they make such great companion plants is a great excuse to grow even more of them.

Here are some of the best herbs you can grow and how to use them as companion plants.

Anise

Anise can help with pest control.

SABENCIA Guillermo César Ruiz, CC BY-SA 4.0 <https://creativecommons.org/licenses/by-sa/4.0>, via Wikimedia Commons: https://commons.wikimedia.org/wiki/File:Pimpinella_anisum._An%C3%ADs.jpg

Scientific Name: Pimpinella anisum

One of the more unusual herbs, anise can grow up to three feet tall and produce white lacy flowers.

Anise is excellent for pest control, repelling aphids and biting insects, and attracting beneficial insects and predators, such as predatory wasps.

Good Companions:
- **Cilantro:** they help each other geminate and produce healthy growth.
- **Brassicas:** anise uses its smell to camouflage these plants and keep pests away.

Bad Companions:
- **Basil, Beans, and Rue:** none of these grow well with anise as it stunts their growth.

Basil

Scientific Name: Ocimum basilicum

Another favorite among gardeners, basil is an excellent companion plant, grows easily in a warm sunny garden, and suits greenhouse growing. Be sure to water basil often, or it can wane and die.

Good Companions:
- **Tomatoes:** both plants improve each other's flavor.
- **Chamomile:** helps the basil grow fast and strong and increases the oil in its leaves.

Other plants you can pair with basil are:
- Chili peppers
- Asparagus
- Beetroot
- Beans
- Bell peppers
- Cabbage
- Eggplant
- Potatoes
- Oregano
- Marigolds

Bad Companions:

- **Sage and Rue:** both stunt the growth of basil.

If basil is allowed to flower, it will attract many beneficial insects to the garden. It also repels many pests, including mosquitos, hornworms, aphids, asparagus beetles, and whiteflies.

Borage

Scientific Name: Borago officinalis

Borage is a popular companion plant, mostly because it attracts pollinators and beneficial predators to the garden.

Good Companions:

- **Tomatoes and Cabbages:** repels cabbage and tomato worms, which can decimate your crops.
- **Strawberries:** helps improve the flavor of the strawberries.
- **Basil:** borage attracts pollinators and good pollinators, while basil repels insects, protecting each other. Borage also improves the flavor of basil.
- **Beans and Peas:** borage loves the extra nitrogen from the beans, returning the favor by attracting good insects and repelling the bad.
- **Cucumber, Melons, Grapes, Peppers, and Eggplant:** Borage feeds the soils with calcium and potassium, bringing the right pollinators in and repelling the pests.
- **Marigolds:** Borage grows better near marigolds; together, they are a pest-repellent powerhouse.

Bad Companions:

- **Potatoes:** if your potatoes have blight, it can kill the borage.
- **Fennel:** at best, it will stunt the growth. At worst, it will kill the borage.

Catnip

Catnip can attract cats.
D. Gordon E. Robertson, CC BY-SA 3.0 <https://creativecommons.org/licenses/by-sa/3.0>, via Wikimedia Commons: https://commons.wikimedia.org/wiki/File:Catnip_flowers.jpg

Scientific Name: Nepeta cataria

Most people know catnip for its ability to attract cats, but it is also a type of mint. Because it attracts cats, you should cage it when you plant it near vegetables, or the cats may destroy them.

Good Companions:

- **Beans:** catnip deters Japanese beetles
- **Beets, Carrots, and Brassicas:** it also repels the flea beetles that attack these plants.
- **Lettuce:** catnip repels slugs.
- **Strawberries:** it deters many of the pests that attack strawberries.
- **Tomatoes:** benefits from the pollinators that come after the catnip flowers.
- **Squash:** any member of this family benefits because catnip can repel squash bugs.

Bad Companions:

- **Parsley:** doesn't like mint, and catnip is part of the mint family.

Chervil

Scientific Name: Anthriscus cerefolium

Chervil, also known as French parsley, is popular in France, Spain, and other Western European countries. Its leaves a taste of tarragon, parsley, anise, and licorice, and you can even eat the flowers. It grows up to two feet tall, so be careful where you plant it.

Good Companions:
- **Broccoli:** chervil improves the flavor
- **Radishes:** chervil makes the radishes crisper and hotter
- **Lettuce:** chervil enhances growth and flavor and deters ants and aphids
- **Yarrow:** enhances the essential oils in chervil
- **Alliums:** leeks, onions, and chives all help keep carrot flies, among other pests, away from the chervil

Chervil is also a trap crop and will work well when planted near vegetables that attract aphids; chervil attracts the aphids and helps protect the other plants.

Bad Companions:
- **Fennel:** fennel attracts aphids that can damage chervil and change the taste of the chervil.
- **Mint:** stunts the growth of the chervil plants.
- **Dill:** attracts pests that destroy chervil.
- **Cilantro:** similar soil requirements and will compete for nutrients.

Cilantro

Scientific Name: Coriandrum sativum

A popular kitchen herb, cilantro is also known as coriander and is a great companion to many other plants.

Good Companions:
- **Beans and Peas:** because they fix nitrogen into the soil, they help feed the cilantro. They also increase the good microbes in the soil, which helps with nutrient uptake. Peas and pole beans also provide shade.

- **Leafy Greens:** cilantro attracts good insects that benefit the greens by feeding on their pests.
- **Tall flowers:** any tall flower will work because they provide a windbreak for the cilantro and shade and act as trap crops for the pests.
- **Anise:** cilantro helps anise germinate quicker.
- **Basil and Parsley** require the same growing environment, so they are easier to care for.
- **Tomatoes:** provide shade, but don't plant them too close together as cilantro needs a lot of nitrogen, and tomatoes don't.
- **Potatoes and Eggplants:** cilantro acts as a trap crop, attracting the Colorado beetle that can destroy your plants.
- **Asparagus:** cilantro deters the asparagus beetle and improves asparagus growth.

Bad Companions:
- **Dill:** they stunt each other's growth and may cross-pollinate.
- **Fennel:** stunts growth, and they compete for nutrients.
- **Rosemary, Thyme, and Lavender** require more sun and dry soil, whereas cilantro needs less sun and moist soil.

Dill

Scientific Name: Anethum graveolens

Sadly, this herb is no longer grown as much as it used to be, but it is a wonderful companion for many plants. Be aware that it matures in just 90 days, so if you want a constant supply for the kitchen and companionship for your plants for the long term, you'll need to plant some every few weeks.

Good Companions:
- **Brassicas:** repels loopers and cabbage worms and improves plant health.
- **Alliums:** these keep aphids away from the dill.
- **Lettuce:** dill repels pests that attack lettuce.
- **Asparagus:** dill attracts beneficial insects, such as lacewings and ladybirds, to protect the asparagus.

Bad Companions:
- **Carrots:** or any other member of the umbellifer family, like parsnips as dill stunt growth, attracts carrot flies, and you risk cross-pollination.
- **Cilantro:** risk of cross-pollination as they are the same family.
- **Tomatoes:** while dill attracts parasitic wasps that feed on tomato hornworms, it can stunt the plant's growth. You can plant young dill near the tomatoes to improve growth, but pull it before it matures.

Fennel

Scientific Name: Foeniculum vulgare

Fennel is probably one of the most anti-social plants in the garden and isn't a great companion for many plants; most gardeners plant it well away from their main crops. However, it attracts many useful insects, including pollinators, parasitic wasps, ladybirds, and hoverflies. That said, some gardeners do report good results when growing fennel near other plants, so it's a case of try it and see.

Good Companions:
- **Dill:** fennel can improve growth and dill seed production, but there is a risk of cross-pollination.
- **Lemons:** fennel keeps slugs and snails away.
- **Lettuce:** fennel deters many insects, including those that attack lettuce.
- **Mint:** mint and fennel are both invasive species, so they compete for space. This results in both plants being slowed down.
- **Peas:** fennel repels pests and improves growth.

Bad Companions:

The list is too long; fennel tends to stunt growth and is an incredibly invasive plant, crowding out others.

Lemon Balm

Scientific Name: Melissa officinalis

Part of the mint family, lemon balm is an invasive species if allowed to grow unchecked – it is best planted in containers and placed where you need it.

Good Companions:
- **Melons and Squash:** Planting lemon balm in the ground can act as a natural mulch for the melons and squash, attracting plenty of beneficial insects and predators.
- **Beets:** the beets help the lemon balm grow, and the lemon balm attracts beneficial predators to protect the beets.
- **Peas:** lemon balm benefits from the nitrogen in the soil.
- **Brassicas and Tomatoes:** lemon balm attracts the beneficial predators needed to keep the brassicas and tomatoes clear of pests.
- **Radishes:** lemon balm protects the radishes from pests like snails, maggots, and aphids.
- **Carrots:** these don't compete for space, and the lemon balm protects the carrots from carrot flies and other pests.
- **Fruit trees:** lemon balm planted around the base can act as a mulch.
- **Lettuce:** lemon balm deters the pests that prey on lettuce.

Bad Companions:
- **Lavender and Rosemary:** they like different soil conditions - lemon balm likes wet soil, and lavender and rosemary like it dry; planting them together ensures one will die.
- **Fennel:** it stunts the growth of the lemon balm.

Marjoram

Scientific name: Origanum majorana

Marjoram is a wonderful herb as it attracts a lot of pollinators, making it an excellent companion for many plants.

Good Companions:
- **Squash, Zucchini, and Pumpkins:** marjoram improves the taste of all these while deterring pests.
- **Corn:** marjoram repels pests that attack corn.
- **Eggplant:** marjoram deters aphids and spider mites from destroying the eggplant fruit.
- **Onion:** marjoram improves the taste.

- **Peas:** marjoram benefits from the nitrogen in the soil and acts as a living mulch; it also attracts beneficial pollinators.
- **Potatoes:** marjoram helps keep many potato pests and diseases at bay.
- **Nettles:** if you are growing marjoram for its oil, nettles will improve production, and the nettle leaves make great liquid compost.

Bad Companions:
- **Fennel:** fennel stunts growth
- **Tomatoes:** while marjoram deters pests from the tomatoes, it requires less water than tomatoes.

Mint

Scientific Name: Mentha

Mint is an invasive plant.
https://unsplash.com/photos/boadZKqd1YM?utm_source=unsplash&utm_medium=referral&utm_content=creditShareLink

While it is incredibly invasive, mint is also an excellent herb for the garden. It comes in different varieties, including spearmint, orange, apple, chocolate, pineapple, and more.

Good Companions:
- **Kale:** mint deters aphids, cabbage moths, and other pests from the kale.
- **Brassicas:** mint attracts the predators that prey on the pests attacking the brassicas.
- **Peppers and Tomatoes:** mint repels many pests and attracts beneficial insects to the plants.
- **Eggplants:** this combination helps build a light soil structure with plenty of nutrients.
- **Roses:** mint can act as a living mulch – *but be aware it is invasive.* It also keeps the soil aerated and moist.
- **Beets:** the mint deters the pests that attack the beets beneath the ground.
- **Carrots:** mint deters aphids and carrot flies.

Bad Companions:
- **Lavender and Rosemary:** mint needs moist soil, while lavender and rosemary like it dry

Most plants will get on fine with mint so long as you grow invasive species in pots. Otherwise, they will crowd each other out, and nothing will grow properly. If you plant mint in pots, it can be placed anywhere in the garden, acting as an excellent pest repellent.

Oregano

Scientific Name: Origanum vulgare

A popular herb, oregano makes a good companion for most vegetables.

Good Companions:
- **Parsley:** stops oregano from spreading too much and improves its flavor.
- **Tarragon:** this repels the pests that would otherwise attack the oregano and provides the nutrients the oregano needs to thrive.
- **Chives:** they enhance each other's flavors. Chives also repel pests from the oregano, and oregano protects the chives from too much sun.

- **Cucumber:** shades the oregano and stops it from spreading too far, and the oregano keeps cucumber beetles away.
- **Strawberries:** the oregano keeps pests away from the strawberries, and the strawberries provide ground cover.
- **Cabbage:** the oregano keeps pests away from the cabbages, and cabbage provides some shade for the oregano.
- **Watermelons:** watermelons protect the oregano from the sun and provide a climbing place, while the oregano keeps the pests away and brings beneficial insects in.
- **Peppers:** each repels pests from the other and requires the same growing condition; oregano also improves the flavor of the peppers.
- **Beans:** oregano repels pests and enhances the growth of the bean plants, and the beans provide the oregano with nitrogen.
- **Asparagus:** oregano improves flavor and acts as a pest repellent, while asparagus shades the oregano, loosens the soil, and makes drainage better.
- **Tomatoes:** oregano repels tomato pests and is a natural fertilizer.

Bad Companions:
- **Mint:** both have different moisture requirements but are both invasive species.
- **Chives:** compete for the same nutrients; neither will thrive.
- **Basil:** has different moisture requirements; they will grow well together if grown in pots.

Parsley

Scientific Name: Petroselinum crispum

Parsley is easy to grow and has several varieties, all with unique flavors.

Good Companions:
- **Asparagus:** they each improve growth in the other, and parsley repels pests, such as the asparagus beetle.
- **Tomatoes:** parsley attracts hoverflies that attack aphids and acts as a trap crop.
- **Peppers:** parsley deters pests and improves the pepper's flavor.

- **Corn:** parsley repels pests that attack corn and attacks parasitic wasps and other beneficial predators.
- **Chives:** protect the parsley from carrot flies.
- **Basil:** parsley improves its flavor and repels some pests.
- **Beans:** parsley benefits from nitrogen and, in return, acts as a pest repellent.
- **Brassicas:** parsley deters cabbage worms from attacking your crop.
- **Roses:** parsley protects roses from aphids and many other pests.
- **Fruit trees:** parsley repels the codling moth, gypsy moth, and other pests that attack fruit trees.

Bad Companions:
- **Mint:** too invasive and will crowd out the parsley unless planted in pots, and will also affect the flavor of the parsley.
- **Carrots:** both want the same nutrients from the soil, and both attract the same pests.
- **Lettuce:** parsley hastens bolting.
- **Alliums:** can stunt parsley's growth and affect its flavor.

Rosemary

Scientific Name: Rosmarinus officinalis

Rosemary is a popular herb and is relatively easy to grow. However, if planted in the ground, it needs regular pruning to stop it from growing too leggy.

Good Companions:
- **Brassicas:** rosemary masks the smell from the brassicas, helping confuse and deter pests.
- **Beans:** feed the nitrogen into the soil for the rosemary and provide shade. In return, the rosemary deters the Mexican bean beetle and improves the bean plant's health.
- **Carrots:** rosemary keeps pests away from the carrots and helps improve growth and flavor. In return, carrots feed the soil and improve soil structure, thus improving healthy growth in the rosemary.
- **Marigolds:** keep pests away from the rosemary.

- **Strawberries:** rosemary keeps pests away from the strawberries, and both improve each other's growth. Rosemary also improves fruit flavor.
- **Peppers:** rosemary keeps the pests away and acts as ground cover, keeping the soil moist and the weeds down.
- **Onions:** they both repel pests from each other, and rosemary makes onions taste better.
- **Parsnips:** rosemary keeps the carrot flies at bay.

Bad Companions:
- **Basil:** needs more water than rosemary.
- **Mint:** both are invasive and unless planted in containers, neither will thrive.
- **Tomatoes:** these require more water than rosemary, and rosemary can inhibit tomato growth.
- **Pumpkins:** both are prone to mildew.
- **Cucumbers:** need more water than rosemary and more nitrogen, which the rosemary can't tolerate.

Sage

Scientific Name: Salvia officinalis

Sage attracts pollinators.
https://unsplash.com/photos/3tGxWVuSRBk?utm_source=unsplash&utm_medium=referral&utm_content=creditShareLink

Sage isn't grown so much these days, but it is a wonderful herb for attracting pollinators and is easy to grow.

Good Companions:
- **Brassicas:** sage is a pest repellent and helps to protect brassicas from cabbage worms, cabbage moths, and more.
- **Carrots:** sage deters carrot flies.
- **Strawberries:** sage keeps the pests away and improves strawberry flavor.
- **Tomatoes:** sage keeps the pests away and brings the beneficial insects in.

Bad Companions:
- **Alliums:** need more moisture than sage.
- **Cucumbers:** sage stunts their growth and makes the cucumbers taste bitter.
- **Rue:** inhibits sage growth.

Tarragon

Scientific Name: Artemisia dracunculus

Tarragon is not seen much in gardens today but is an excellent companion for vegetables. When you buy seeds, ensure they are not the Tagetes lucida variety; this is a tarragon substitute and not as strong.

Good Companions:
- **Chives:** these deter pests from the tarragon, and both like the same soi and moisture conditions.
- **Eggplant:** like the same moisture levels, and tarragon improves fruit flavor.
- **Cilantro:** both like the same growing conditions, and cilantro keeps spider mites away from the tarragon.
- **Garlic:** protects tarragon against spider mites, and tarragon improves garlic growth.

Bad Companions:
- **Sage, Oregano, Rosemary, Lavender:** all these prefer drier soil and won't thrive if you plant the moisture-loving tarragon with them.

Thyme

Scientific Name: Thymus vulgaris

Thyme is very simple to grow and comes in several varieties. It is a magnet for pollinators.

Good Companions:

- **Brassicas:** the thyme repels cabbage worms, beetles, slugs, and cabbage loopers. It also attracts beneficial predators, such as ladybirds.
- **Tomatoes:** thyme repels tomato hornworms and improves tomato growth.
- **Potatoes:** thyme attracts beneficial predators that keep potato pests at bay and improves the potato's flavor.
- **Eggplants:** thyme repels hornworms, aphids, beetles, spider mites, and garden moths.
- **Strawberry:** thyme repels pests and attracts beneficial pollinators.

Bad Companions:

- **Most Herbs:** need different soil and moisture conditions.

In terms of companion plants, we'll move on to flowers next before giving you a look at how to plant as a pest repellent.

Chapter 6: Companion Planting with Flowers

You now understand that companion planting is one of the best natural ways to protect your plants from pests and diseases, among other benefits. We looked at vegetables and herbs, but what about the humble flower?

Flowers are excellent for adding to a vegetable garden. They provide a wonderful pop of color and attract the best beneficial insects and predators to protect your plants. However, there are some things you need to consider when using flowers as companion plants.

- **Blooming Time:** If you want pollinators to come into your garden, you need to be sure that the flowers bloom when your flowering vegetables do; otherwise, the vegetable flowers won't be pollinated.
- **Growing Conditions:** flowers must be grown with vegetables that like the same growing conditions, i.e., soil type, water, and light. You also shouldn't grow tall flowers that shade your vegetables all day.

Flower Categories

There are three categories of flowers, each with their own defining characteristics:

1. Annuals
2. Biennials

3. Perennials

Let's look at these in more detail so you understand what to plant and when:

Annuals

Annuals take one year to complete a life cycle; this means they grow from seed and flower and produce seed in one growing season. You can also get hardy annuals; these can be planted in the fall and will sprout in the spring.

Biennials:

Bi means two, giving a clue to the fact that these types of plants and flowers complete a lifecycle every two years. The first year is purely aesthetic, and the second is when the seeds are produced.

Perennials:

Perennials are hardier than the others and will die back after the growing season, ready to regrow the following year. These are good for attracting beneficial insects and predators.

What you grow depends on your zone, your vegetable choices, and which flowers are your favorites. Don't forget that most herbs flower, too, so you get the best of both worlds. However, this chapter will focus only on actual flowers, so let's look at some of the best for companion planting.

Calendula

Also called the pot marigolds, these are different from other marigolds. Calendulas are simple to grow and flower throughout the season so long as you deadhead them. You can also harvest the seeds, dry them, and use them the following year.

Calendulas are a pest repellent, deterring asparagus beetles and tomato hornworms. They also make an excellent trap crop as they attract aphids away from other plants.

Chamomile

Chamomile is bright and attractive, bringing in bees and other beneficial insects (including predators that will rid your garden of nasty pests). And, of course, it makes a delicious tea. You can also use cold chamomile tea as a spray on your seedlings to prevent a fungal disease called "damping off."

Chamomile can be dug back into the earth at the end of the growing season to feed it with potassium, magnesium, and calcium. You can also trim it regularly and spread the cuttings at the base of any plant to act as a mulch and feed nutrients into the soil. Chamomile also helps repel mosquitoes.

Comfrey

Wilted comfrey leaves can be used to feed other plants.

Treated by many as a weed, comfrey is actually a very useful plant. It does have a very long root system. This is brittle, and should you leave even the tiniest bit in the soil, it will grow back. Plant it in containers if you don't want it to invade your garden. Bees are attracted to comfrey flowers, and the leaves can be used to make compost tea or as mulch. You can also layer wilted leaves at the bottom of a potato trench to feed the potato plants with phosphorus, nitrogen, and potassium.

Cosmos

An easy-to-grow annual, cosmos produces an excellent, colorful display. The white and orange varieties attract pollinators and green lacewings, one of the best predators that feed on aphids, thrips, and other soft-bodied insects that destroy vegetables.

Marigolds

Common in veggie gardens the world over, marigolds are excellent at attracting beneficial insects. They grow just about anywhere and come in a few varieties. The most useful is the French marigold, which helps to deter nematodes in the soil by producing a chemical from their roots. However, it can take a couple of years for this chemical to build up sufficiently, so be patient. When the season is over, chop the plant back, but leave the roots intact. Dig the foliage into the soil, and it will break down.

Because of this quality, they are good companions for most vegetables and fruits. They also deter rabbits when you plant them around your garden. Planting them with beans helps repel the Mexican bean beetle, while they can also repel squash bugs, tomato hornworms, whiteflies, and thrips – an all-around excellent companion plant.

Nasturtiums

The perfect trap crop, the brightly colored flowers on the nasturtium attract aphids and blackflies, tempting them away from valuable vegetables they would otherwise destroy. Once infested, simply remove the nasturtium stems and destroy them. They repel many types of beetles and bugs. As a bonus, the leaves and flowers are edible.

Roses

Roses are not considered good companion plants.
https://unsplash.com/photos/dv7cSiHurKM

Roses are not great companion plants because they tend to attract a lot of pests. However, that makes them good as a trap crop when planted away from your vegetables and fruits. They are especially good at attracting aphids away from grapes. You can protect the roses from some pests by planting garlic nearby and garlic chives; the latter will also flower and attract pollinators.

Sweetpeas

Another colorful crop, there are plenty of sweetpea varieties to choose from. These work well with pole beans to attract pollinators.

Let's bring this together by looking briefly at companion planting for pest control.

Chapter 7: Companion Planting for Pest Control

As you've discovered over the last few chapters, some plants are great for deterring pests by using strong aromas to mask smells from other plants or attracting beneficial insects, pollinators, and predators that feed on pests.

Pests are undoubtedly one of the biggest problems gardeners face, and while chemicals can keep them at bay, they cause serious, sometimes irreversible damage to the ecosystem in your garden. Organic is the way forward, and companion planting is the best way.

Pests are one of the largest problems gardeners face.
https://unsplash.com/photos/8oT2MA33jsk?utm_source=unsplash&utm_medium=referral&utm_content=creditShareLink

Different plants work in different ways, and while companion plants can help with pests, some do need time to build up protection to sufficient levels. For example, marigolds need a couple of years to build up natural chemical levels in the soil to deter nematodes.

It's also worth noting that companion planting won't provide a complete solution once a pest infestation has occurred. Take onions and cabbages, for example. If you plant the onions first, they will protect your cabbages from the dreaded cabbage moth. It's advisable to plant the onions last. However, if the cabbage moth has already attacked, the onions won't do much.

Some years will be worse than others for pest infestations. The insect populations ebb and flow like the tides, and some years, you won't have much trouble at all, while other years, you might just wonder why you bothered planting a garden at all! It also depends on what is growing nearby. If your neighbor has a garden full of plants that attract pests, you'll likely have many of them in your garden, too. On the other hand, if they grow a garden full of plants that attract good insects, you will benefit, too.

It's important to note that companion planting is not a complete solution. While it will deter many pests, you must still keep a close eye on your plants and take action should infestations occur. That means using organic pest controls where needed or hand-picking insects off plants.

Beneficial Insects

Some plants deter pests, but others attract the beneficial ones that feed on the pests. Provide those insects with a good environment and home to thrive in, and you can see an increase in their population above that of the pests. The chart below shows the best insects to attract and their benefits:

INSECT	BENEFITS
Parasitic Wasps	Feed on aphids, grubs, and caterpillars
Lacewing Larvae	Feed on aphids
Ground Beetles	Feed on lots of different ground pests

INSECT	BENEFITS
Hoverflies	Feed on caterpillars, leafhoppers, and many other insects
Ladybird Larvae	Feed on aphids
Robber Flies	Feed on caterpillars, leafhoppers, and many other insects
Pollinators	To pollinate your plants for a successful crop

    A key aspect of using companion planting to control pests is diversity. Don't plant onions and garlic everywhere in the hope that they will keep all the nasties away. You need a good variety of plants to attract the right insects and control the bad ones for them to work effectively.

    Plants must also attract these beneficial insects throughout the growing season, not just a part of it. If your flowers bloom in June but are gone a month later, they won't protect your plants for the rest of the season. Succession planting works to a degree, but diversity is the real key. This is one of the biggest mistakes gardeners make, so ensure your companion plants flower and attract these insects right through the season to provide full protection for your vegetables and fruits. If not, the good insects will go elsewhere to find food!

    An important thing to remember is that beneficial insects won't feed all the time, even though food may be in plentiful supply. At times during their life cycle, they don't feed but do need somewhere to live and survive. Providing everything these insects need throughout their entire life will entice them to live in your garden and work for you, protecting your plants for longer. Fallen logs and hedgerows are good options, or you can provide insect hotels where they can overwinter safely. This also provides ground cover that locks in heat and makes it easier for early flowers to grow (also providing food and nutrients).

    If you want a sustainable hedgerow, consider growing one of the fruit bushes or dwarf/patio fruit trees.

What Do Beneficial Insects Need?

To keep these beneficial insects coming in, there are certain types of plants you need to grow in your garden:

- **Ground Cover:** plants that spread across the ground, like thyme, oregano, rosemary, and sage, will provide cover for all sorts of insects, especially ground beetles. If they can hide from their predators, they can keep working for you and your garden.
- **Shade:** many insects need protected shady areas to lay their eggs.
- **Tiny Flowers:** many pollinators and beneficial predators prefer smaller flowers, like those you get on many herbs. For example, parasitic wasps are tiny and prefer clover, fennel, cilantro, dill, thyme, and so on because their flowers are tiny.
- **Composite Flowers:** other insects prefer larger flowers, like marigolds, daisies, and chamomile, including hoverflies and predatory wasps. Mint plants are good for these, too.

Deterring Pests

Herbs are among the best companion plants. Not only can you use them in your kitchen, but they deter many pests, too. These herbs also look and smell fantastic in your garden, working with the environment and each other to provide a wall of protection. Be experimental and try new things to see what works for your garden and what doesn't.

Common Pests and Companion Plants

Here's a chart showing the common garden pests and the plants that help repel them:

PESTS	DETERRENT PLANTS
Ants	Mint Catnip Wormwood Tansy
Aphids	Chives Catnip

PESTS	DETERRENT PLANTS
	Cilantro
	Eucalyptus
	Chrysanthemum
	Fennel
	Marigold
	Garlic
	Mustard
	Mint
	Oregano
	Onion
	Nasturtium
	Feverfew
Asparagus Beetle	Pot marigold
	Basil
	Parsley
	Tomato
	Tansy
	Nasturtium
Bean Beetle	Nasturtium
	Summer savory
	Rosemary
	Marigold
Black Flea Beetle	Sage
Cabbage Looper	Eucalyptus
	Hyssop
	Dill
	Garlic

PESTS	DETERRENT PLANTS
	Peppermint or spearmint Onion Nasturtiums Thyme Sage Wormwood
Cabbage Maggot	Radishes Marigold Garlic Wormwood Sage
Cabbage Moth	Mint Sage Rosemary Hyssop Tansy Summer savory Thyme Oregano
Cabbage Worm	Tomatoes Thyme Celery
Carrot Fly	Alliums Rosemary Lettuce Wormwood Sage

PESTS	DETERRENT PLANTS
Colorado Potato Beetle	Cilantro Marigold Onions Nasturtiums Catnip Eucalyptus Tansy
Corn Earworm	Geranium Cosmos Marigold
Cucumber Beetle	Marigold Radishes Catnip Nasturtium Tansy
Flea Beetle	Garlic Rue Wormwood Tansy Sage Mint Garlic Catnip – steep leaves in water and spray
Flies	Tansy Rue Basil

PESTS	DETERRENT PLANTS
Japanese Beetle	Chives Hydrangea Tansy Rue Pansy Garlic Catnip
Leafhopper	Geranium Chrysanthemum Petunias
Mexican Bean Beetle	Petunias Summer savory Rosemary Marigold
Mice	Wormwood Tansy
Mosquitos	Garlic Rosemary Geranium Basil
Moths	Rosemary Lavender Wormwood
Peach Borer	Garlic
Nematodes	Calendula

PESTS	DETERRENT PLANTS
	Marigolds – it takes around 1 year for chemical levels in the soil to build
Onion Fly	Garlic
Snails and Slugs	Garlic Fennel Sage Rosemary
Spider Mites	Cilantro
Squash Bugs	Mint Catnip Nasturtiums Radishes Petunias Tansy
Squash Vine Borer	Radishes
Striped Pumpkin Beetle	Nasturtiums
Ticks	Lavender Garlic
Tomato Hornworms	Calendula Borage Marigold Dill Petunias

PESTS	DETERRENT PLANTS
White Cabbage Moth	Mint
Whitefly	Marigold Basil Thyme Peppermint Oregano

When planting for pest control, make sure you choose plants that protect throughout the whole growing season. However, you will still need to monitor your plants as there is no one-size-fits-all solution.

To finish Part Two, we'll look at whether you should use seeds or starter plants.

Chapter 8: Seeds vs. Starters

Do you start with seeds or starters/seedlings from your local nursery? This is a big decision, and you need to base your decision on several factors. For the most part, you'll use a mixture of both. Some of the things you need to consider are:

- Maturing time
- Size
- Transplantability

Let's examine these features in greater detail.

Maturity Time

Different plants take different times to reach maturity. For example, tomatoes and peppers are best grown from seedlings rather than seeds, as they can take a long time to come to fruition. If your growing season is short, you won't get any joy by starting these from seed unless you start them indoors early in the year.

In contrast, spinach and lettuce don't take long to mature. Harvest can often be achieved within 30 days of planting the seeds.

Spinach plants can be harvested within 30 days of planting the seed.
https://unsplash.com/photos/4VMqrwYfmDw?utm_source=unsplash&utm_medium=referral&utm_content=creditShareLink

Knowing the time a plant takes to mature is key to knowing whether to plant seeds or not. Typically, the faster-growing plants can be started from seed, while slow-growers are best as seedlings from a nursery.

Check the instructions on the packet to discover how long they take to grow and if they need to be started indoors. Purchasing seedlings can often make more sense if the seeds take too long to grow.

Tip

Before deciding, check the maturity time. More than 65 days, you can purchase seedlings, while less than 65 days indicates you can grow from seed. However, as you will see shortly, there are exceptions to this rule.

Plant Size

Plant size is also an indicator. Typically, the bigger the plant, the longer it will take to mature and the longer it takes before you can harvest it. These are best planted as seedlings.

In contrast, smaller plants have a shorter maturity time and can be grown from seeds. Some of these, like radishes and spinach, can be sown directly into the ground, while others are best grown in pots and transplanted when they are big enough.

Transplantability

Some plants are unhappy being moved once they have grown from the seed, as their roots are somewhat fragile. For example, legumes like beans and peas are unlikely to thrive if you try to transplant them. Although they take two to three months to mature, they are best seeded directly into the ground or pre-sprouted – that means laying them on a piece of paper towel, covering them with another, and placing them all in a plastic bag. Spritz them with water to keep them damp, and once they have sprouted, you can put them in the ground.

Other plants that take a while to grow but hate being moved are zucchini, squash, and cucumbers.

Some smaller, fast-growing plants also don't take well to being moved. These include lettuce, arugula, and other small leafy greens. You should seed them directly into the soil or purchase seedlings from the nursery.

Lastly, root vegetables don't like being moved either – beets, carrots, potatoes, etc. Not only are their roots sensitive, but they also feed on nutrients in the soil, and moving them will stop that in its tracks.

Plants to Buy from the Nursery

If you are new to gardening or simply don't have the time or interest to grow from the seed, there are some plants that you should purchase from a nursery to get your growing season started quickly.

Chives

Garlic, chives, and onions will come back every year, regardless of climate, and they can be divided – more plants for your money. However, normal chives are a little tough to grow from seed but are one of the best repellents you can have.

Large Brassicas

Cauliflower, cabbage, Brussels sprouts, mustard, kale, and collards are all large plants that take a long time to mature. These are well worth purchasing from a nursery to give you a jump start on the season. However, if you have the time and space, try growing them from seed – you'll need to start them indoors early in the year. You might just find that they taste better and are healthier plants.

Nightshades

Eggplants, peppers, and tomatoes are tough to grow from seed without a long, warm growing season. All three take a long time to grow from seed, so make sure to purchase and start the seeds early in the season.

Perennial Herbs

Thyme, rosemary, oregano, tarragon, and sage all take a while to grow from seed, and many nurseries typically take cuttings from a healthy plant to grow new ones - you can do this, too once your herbs are fully grown. Provided the nursery plants are healthy, there's nothing wrong with bringing them home and planting them. You can also buy some herb plants from grocery stores.

Swiss Chard

Swiss chard is biennial, and it will last for two years. Young plants are usually readily available in nurseries, and you can enjoy them for a couple of years.

Purchasing Seeds vs. Plants

Seeds

Buying seeds allows you to control the growing environment.
Photo by Zoe Schaeffer on Unsplash https://unsplash.com/photos/silver-glittered-heart-on-persons-hand-XuJreNkw2BM

Buying seeds has an advantage – one small, cheap packet usually has plenty of seeds. Not only that, but if you grow from seed, you get to control the growing environment, and you know that your plants grow in organic, nutrient-rich soils – you don't get that with nursery plants. However, make sure you purchase seeds from companies that prioritize non-GMO, organic seeds.

Plants

Plants are more expensive, but there will be occasions when it's better than buying seeds. Ensure you buy from local nurseries or organic growers rather than the big-box stores. Store plants have usually traveled a long way to get to the store and have likely been sprayed with chemicals to keep them fresh for longer. They've also likely been fed chemical fertilizers. If you try to feed them something different, they won't like it and won't grow properly. Plants like to be treated exactly the same throughout their entire growing cycle; try to change something, and they'll go into a sulk, which could lead to death.

Tips for Buying Healthy Plants

When you need to buy plants, there are a couple of things you should do to make sure you get a strong, healthy plant:

- Choose plants that have not started blooming. Get smaller plants. These can spend time growing their root system when you plant them, rather than flowers and fruit. If you can't help but purchase flowering plants, pinch off the flowers when you transplant them.
- Remove the plant and soil from the pot and look at the roots. If it spirals around the plant, it tells you it has been in the pot for a long time and is rootbound. Plants like this will struggle to establish themselves in your garden. Make sure the roots look healthy and white.
- Look for diseases and pests on the leaves, paying close attention to the underside and the stem.
- Do not buy plants that have gone leggy, tall, and narrow. They have not had enough light or have been grown in overcrowded conditions, which means they are already stressed and unlikely to thrive.
- Buy plants certified as organic. If you can't, look for labels that say they are naturally grown. Many smaller organic growers don't

have the money to become approved but don't use chemicals on their plants.
- Use local nurseries. You can ask them questions about their plants and are more likely to find organic ones. Not only that, but you support a local business rather than a big-box store. These nurseries will also have plants grown in your locale, which means you know they'll be okay in your climate.

Let's move on to the real work – growing and caring for your plants.

SECTION THREE: PLANTING, CARE AND MAINTENANCE

Chapter 9: Start with the Soil

Healthy plants require healthy soil; it's as simple as that. If your soil is healthy, you don't need to use so much fertilizer. Instead, the soil is filled with rich organic materials, such as decaying grass clippings, leaves, and compost. It should retain moisture but have good drainage, be loose and full of the air the plants need for their roots, and be filled with minerals to help them grow. It will be populated with living organisms that help maintain its quality.

Healthy soil will help you grow healthy plants.
https://unsplash.com/photos/fjj7lVpCxRE?utm_source=unsplash&utm_medium=referral&utm_content=creditShareLink

If your plants are happy, look at the soil first. Before we dig deep – pardon the pun – into soil health, here's a quick fix to improve your soil.

Quick Fix

As a beginner, soil health can be overwhelming. This quick fix can help you with your soil before you put your plants in:

1. **Clear the Debris:** clear away rocks, stones, and other debris. If you need to remove grass, use a sharp spade to cut it into smaller, easier-to-handle pieces first.
2. **Loosen the Soil:** if your garden hasn't been dug before, use a good spade and fork to loosen the soil. You need to go 8-12 inches down to give the roots room to grow.
3. **Add Some Organic Matter:** add aged manure and compost to feed nutrients into the soil and improve drainage. This will create pockets in the soil where oxygen can enter and mix. You need 2-4 inches – no more, no less – spread over your soil and dug in. If your garden is already well dug, simply layer the compost on top and leave it – the worms will dig it in for you.

Digging Deeper

Do you know if your soil is sandy or clay? Alkaline or acid? Nutrient-rich or poor? Those are the things you need to know to be a successful gardener. It's the only way to know if you need to make changes to make things grow better, and we're going to look into all three of those in more detail.

Soil Type

There are three basic soil types: sandy, clay, and silt.

What you are looking for is an equal mix of the three. This gives good water retention and drainage with enough space for oxygen to mix. It is also light enough to make it easier to move and manipulate. But how can you tell if you have loamy soil?

It should not be sticky, even after rain. It should be damp, crumble easily, and not crust over when dry. In contrast, clay soil is sticky, retains shape when crushed, and doesn't drain well. It cracks, goes solid in the summer, and is waterlogged in fall and winter. Sandy soil is gritty, loose, and won't hold its shape. It drains too quickly, loses nutrients, and requires amending with manure and compost.

Testing Your Soil

You could opt for an official soil test, which may not be cheap. Alternatively, use a simple DIY jar test:

1. Get several Mason jars
2. Pick several sits from your garden and dig down around 6 inches – root level for many plants.
3. Take a sample of the soil and half-fill each jar, labeling the jars with each area of the garden.
4. Fill each jar with water and leave it to one side. Once the water has soaked in, pop on the life and agitate each jar for 2-4 minutes.
5. Allow the contents to settle, and measure the layer of sediment found at the bottom of the jar. This is your sand content.
6. Do the same again after letting the mixture sit for 4 minutes. Take this measurement and subtract your previous measurement. This is your silt content.
7. Allow the jar to sit for a full day, and repeat the process, subtracting your second measurement (not the second minus the first). This is your clay content.
8. Add the three, divide each measurement by the total, and multiply by 100 to get the percentages.

You want around 40% sand and silt and 20% clay.

If the percentages are off, you can add nitrogen-rich organics to decrease the sand content.

If your soil is too silty, add coarse grit or small gravel and some compost or aged manure with additional straw.

Add coarse grit, peat moss, and compost if your soil has too much clay.

Soil Nutrition

Soil tests can also tell you how fertile your soil is. Poor fertility will lead to a poor garden. You need the soil to be packed with potassium, phosphorous, and nitrogen to aid in plant growth. When purchasing fertilizer, the letters N, P, and K on the packaging denote these three nutrients.

- **Nitrogen: N** – adds the green to your plants. The leaves and stems will benefit when they have nitrogen. Older manure is usually dense in nitrogen, along with calcium-rich organics, sea life, and blood.

- **Phosphorus: P** – the catalyst. It is great for the early stages of plant growth and will ensure strong roots, which will benefit blossoms and fruit. It can be found in both quick and slow-release.
- **Potassium: K** – the defender. It will protect all parts of the plant, adding resistance and boosting the taste of vegetables and fruits.

While your plants need these three nutrients, too much can do as much harm as too little. Be sure to research how much your plant needs and add only that.

Soil pH

If the pH of the soil is off, the nutrients will not be transferred to the plants. Too low or too high, and your soil may be nutrient deficient or toxic, neither of which suits your plants.

Try to keep the pH of your soil between 6-7. However, some plants can take a range of pH while others need a specific pH level.

There are plants that thrive in acidic soil, such as blueberry bushes, but they are generally the exception to the rule. If your soil is too acidic, garden lime will fix that up. If it is too alkaline, add in some sulfur.

Be aware that this is not a quick fix. It can take upwards of a year for the amendments to have any real effect, but you only need to change your soil pH if your plant won't grow in it.

Common Soil Amendments

- **Plant Material:** most cut organics such as grass, leaves, and all softer materials (not branches, heavy roots, etc.). They need time to decompose and can greenly be composted in the fall for planting in the spring.
- **Compost:** rotting plant materials and organic scraps. Add these to your soil a few weeks before planting to add balance.
- **Leaf Mold:** lots of nutrients are packed into decomposing leaves.
- **Aged Manure:** As it ages, it develops more nutrients and loses much of its acidity. It can be smelly to store, but add it too soon, and you risk adding disease to your soil.
- **Coconut Coir:** conditions the soil and helps it retain moisture. This is more sustainable than using peat moss.

- **Woodchips, Bark, and Sawdust:** compost these before adding them; otherwise, they will steal the nitrogen and starve the plants
- **Green Manure:** commonly called cover crops, they improve the soil. Plant in the fall, and chop and work them in in the spring. They add nutrients and soil structure.
- **Garden Lime:** raises pH
- **Sulfur:** lowers pH
- **Wood Ash:** raises pH

The last 3 should only be used if a soil test recommends them.

Adding Organic Matter

If you add organics in the fall, they have time to break down before the spring growth period. Don't be tempted to add everything at once. The material needs to start decomposing before you add more, and it needs space to do so. Take your time, and the compost will be packed with nutrients.

If you didn't do it in the fall, make sure you do so as early in spring as possible:

1. Spread at least a 2-inch layer of organic matter onto your garden; no more than 4 inches. A garden fork will help aerate the soil as you mix it. Mix the organics with the top half foot of soil, making sure of an even distribution.
2. You need to build over time, so add a little more each year. This allows the nutrients to build up.
3. Once you have your soil and compost, add a generous amount of water.
4. Don't plant in the soil immediately. Leave the soil for 2-3 weeks before you begin planting your garden.
5. Rake off any debris that might have fallen before you begin planting, and ensure the soil is even and level. Now, you can plant!

The compost adds nutrients through microorganisms, but too much of a good thing is never good. If they develop and grow too quickly, they can deplete the nutrients instead of adding to them. This will mess up the pH level of your soil. It needs to comprise about a quarter of your soil mixture and should be mixed into your soil thoroughly.

Using Raised Beds

You could use raised beds if your soil cannot be amended quickly enough. This way, you control the soil and its nutrient levels. Do NOT walk on the soil in your beds; it will compact quickly and go hard. Keep your beds no more than 4 feet wide, or place a pathway down the center if you want wider ones.

Raised beds limit frost and ice in colder areas, and you can plant a few weeks earlier without worrying about damaging the seed pods. You can cover them with a non-porous dark material to keep weeds down and warm up the soil to start growing in them earlier.

Using Cover Crops to Improve Soil

Soil fertility is important, as you know by now, and there are some underlying principles to help you maintain healthy soil:

- Keep the soil covered as much as you can
- Do not disturb the soil unless necessary
- Keep roots growing all year around to feed the soil
- Diversify what you plant

That's where cover cropping comes into its own. Otherwise known as green manure, cover crops help you do all that and are a cost-effective way of maintaining rich, fertile garden beds and improving quality and yield.

If you want to go down the route of adding amendments every season, go right ahead. But cover cropping is a great way to achieve the same results without the heavy lifting. Simply dig your beds in the fall, scatter the seeds over the top, cover them with a thin layer of soil, and water them. Keep them moist until they germinate, and let them grow.

Once a cover crop is established, it will keep the weeds down and stop virtually all topsoil erosion caused by the wind. When the crop is ready, simply chop it, drop it onto the garden, and leave it to die a little before digging it in – including the roots. As it decomposes in the soil, it will feed it with nutrients and nitrogen, improving microbial activity and soil quality. As these crops are planted when your garden would otherwise be empty, they do all the work for you while you can sit back and take a break from the hard work.

The soil is full of living microorganisms, and cover crops feed these. However, it's a symbiotic relationship. Those microorganisms feed the plants with nutrients like phosphorus and nitrogen.

Cover crops with long taproots keep soil compaction to a minimum, thus aiding plant growth. Their roots dig deep down into the soil and aerate it, encouraging moisture to go deeper into it and reducing the runoff risk.

Which cover crops you plant depends on what you want. If your soil is clay, compacted, sandy with little fertility, suffers from soil erosion, or has insufficient organic matter, it will benefit from a cove crop, so first, work out what you need to fix.

Best Cover Crops

Cover crops are usually perennials with a short lifecycle, annuals, or biennials. They all have advantages and disadvantages, and these are some of the best ones to use, with tips on how to get rid of them.

Winter Rye

Scientific Name: Secale cereal

An annual grain, winter rye adapts to all types of soil, even soil with little fertility, sandy, and acidic soils. It is a crucial cover crop for the winter as it suppresses weeds, and its roots exude a chemical that stops weeds from germinating. It is planted as the last crop in the season because it needs cool soil to germinate, typically when the temperatures are down to 35°F. Typically, this means planting it in the fall, about the time of the first frost, any earlier, and it can turn aggressive.

How to Remove It

In the early spring, move the rye and leave it on the ground or till it is straight into the soil. A warning, though- it doesn't die in the winter, so you need to time your removal to stop it from going to seed. If you are mowing and leaving it, leave it a few weeks to decompose before you start planning in the garden.

Field Peas

Scientific Name: Pisum sativum

Peas like cool weather. Plant when the hot summer weather starts or when it has passed. Peas put nitrogen into the soil to feed future crops. Rhizobia are bacteria in the soil that convert atmospheric nitrogen into something plants can use, eventually attaching themselves to pea roots. When the peas have died back, the nitrogen stays in the soil. To plant these, bury each pea up to 2 inches deep in the soil.

How to Remove It

If you plant in the spring, the peas will die by the late spring and can stay where they are to decompose, or you can till them back into the soil before you replant. If you plant in the fall, their stems die in winter and will be fully decomposed before the spring growing season starts.

Oats

Scientific Name: Avena sativa

Oats are cool-season grass that germinates energetically, establishing itself very quickly, especially spring-grown oats. Their profuse top growth and stringy roots improve soil structure when you till them in. They scavenge phosphorus from the soil, and plantlets hoover up extra nutrients, which restore soil fertilization levels. You can plant these in the fall or spring.

How to Remove It:

When spring oats have been growing for 6 to 10 weeks, you can mow them and leave them to decompose or till them in, preferably while they still have green seed heads. Fall oats are planted from the third week of September, allowing them to be established before early winter and soil erosion hits. If you are in a cold zone, they will not survive, so you can easily till them into the ground in the spring.

Crimson Clover

Scientific Name: Trifolium incarnatum

Crimson clover is a beautiful plant that can be grown anytime throughout the growing season. It has a simple taproot that mines the soil and builds up nitrogen. It is great at suppressing weeds and controlling erosion, and if left to flower, it will attract pollinators.

How to Remove it

Once it starts to bud, mow it down and till it in. You can leave it until it flowers, but you must be quick, or it will self-seed and grow everywhere. If you plant it in the winter, it will die back naturally in cold winters.

Ryegrass

Scientific Name: Lolium multiflorum

Ryegrass is readily available in most nurseries for spring or fall planting. A packet of seed will usually be a combination of annual and perennial grass. When you plant annual ryegrass in the fall, it will die back in the winter, and you can till it into the soil in spring. The perennial seed is more difficult as its root system is long. Given the right moisture levels, it is prolific in the cool seasons, so plant in late summer through early fall. Any later than that, and it won't establish properly, especially if a frost hits.

How to Remove It

Mow it down in early spring. It may need more than one mowing if you have the combination seed, and you may need to smother it if you want to kill it off completely. Leave it to decompose for a few weeks before you plant in the soil.

Oilseed Radish

Scientific name: Raphanus sativus

This is the best cover crop for compacted soil. It breaks soil up easily with its long taproot and improves soil drainage significantly. It looks more like a daikon radish than your typical spring variety. If you plant it midsummer, it will grow big enough to create pockets in the soil, allowing better air and water filtration and making it easier for seedlings to root themselves in the sprint.

How to Remove It

It will die off once temperatures drop to 20°F or lower and will be fully decomposed by the spring.

Using Organic Fertilizer

Chemical or synthetic fertilizers are cheap and readily available, so why would you want to use the more expensive organic versions on your soil? There are several reasons why. It's not all about the quick result. Synthetics might work right now, but organic fertilizers keep the soil healthy long-term.

They Work Slowly

Before an organic fertilizer can work, it must be broken down by the soil, ensuring the plants and soil get the right nutrition when needed. Synthetic fertilizers usually result in overfeeding, can burn plants, and do not benefit the soil.

They Improve the Soil

Organic fertilizers and materials improve the texture of the soil, helping it retain moisture and significantly increasing the activity of microorganisms. Not only do they help plants, but they help the soil, too. Synthetic fertilizers strip nutrients from the soil, providing a very poor harvest.

They Are Safe

Not to eat or drink, obviously, as most of them would taste disgusting, even though they are natural. However, they are safe for your garden, the environment, your kids, and the family pets. Conversely, synthetics use fossil fuels to produce them, and runoff often pollutes water sources.

They Are Easy to Use

They couldn't be easier; simply mix, spray, or add to the soil. They benefit your garden in many ways and are as convenient as synthetic fertilizers.

Mixing Organic Fertilizer with Soil

You need to consider a few things before you do it. The fertilizer cannot be added as is and must be watered down first. This ensures the fertilizer is evenly distributed.

You also want to add it slowly. Add a little and mix it thoroughly, starting with a ratio of 1:10 fertilizer to soil. Once the right amount has been added, you can water the soil.

First, decide what you want to use – loads of organic fertilizers are available, or you can make your own. The type of fertilizer will depend on your garden and growing needs. Make sure you follow the package directions carefully and add the fertilizer in the right amounts – do not add too much, as it can cause harm to the soil and plants.

Mix it in, using a garden fork or space to distribute it through the soil evenly.

Lastly, plant your plants and ensure you give them enough room to grow.

In the next chapter, we'll get down to the nitty-gritty of planting your plants and their companions.

Chapter 10: Plant Those Pairs

You know what you want to plant, and you've got your soil just right, so now it's time to get your hands dirty. This chapter will start by looking at planting seeds vs. seedlings.

Seeds

Growing plants from seed might seem daunting to some beginners, but it isn't that difficult. It is far more rewarding than buying plants. Once you get into it, it's fun. This section will give you the confidence to start growing your plants from seed, be they vegetables, herbs, or flowers.

So, why would you grow from seed when you can head to the nursery and pick up everything you need?

As mentioned earlier, some plants are better grown from seeds, particularly those that don't take long to mature. That said, it really is down to you how you grow your garden, but it's a guarantee that once you start growing certain plants from seed, you won't look back.

Don't pressure yourself. If things don't go right, you can always buy the plants from the nursery and have another go at growing from seed. However, to push you in the right direction, here are some of the primary benefits:

- **It's Cheaper:** buying a pack of seeds is far cheaper than buying the plants, and you get more for your money. However, keep in mind that seeds expire, so check the dates on the packets.

- **More Choice:** there will always be more choices with seeds as opposed to plants, so you get more options for your garden and way more varieties of each vegetable, herb, or fruit, too.
- **You Know What You Are Growing:** when you grow from seed, you control the growing environment and the fertilizers, pesticides, and fungicides you use. With any luck, you'll be doing it all organically. You have no idea what chemicals, if any, have been used with nursery plants.
- **You Can Start Earlier:** especially if you live in a colder climate, you get to start your seeds indoors earlier and have the pleasure of watching them grow into healthy seedlings ready for the spring.
- **Pride:** you grew those plants, so you have every right to be proud of yourself for your achievements.
- **Plenty for Everyone:** whenever you grow from seed, you always sow a few extra, just in case some don't make it. That means you'll likely have extras; share them with your friends and family or sell them to make a bit of cash for next year's seeds.

Seed Starting 101

Most beginners struggle because the technical stuff confuses them, or they try to be clever and do things in a complicated way. Growing from seed isn't that difficult, but let's clear up some of the basics for you right now!

Technical Terms:

Yes, gardening does come with a few technical terms, but they aren't difficult to understand; you'll soon be using these words like a pro! These are some of the most important ones:

- **Sowing:** nothing more than planting your seeds
- **Germination:** when your seeds begin to form seedlings
- **Scarification:** scratching a head seed's outer coating to speed the germination process
- **Stratification:** simulating cold-weather conditions for those seeds that need to be dormant in the cold before they can germinate

Seed Starting Techniques

Success at growing from seed depends on you doing it the right way, and you can use two main techniques: indoor and direct sowing.

- **Indoor Sowing:** your seeds are sown into containers and kept indoors for several weeks before transplanting the seedlings into the garden. This way, you can start your crops off much earlier than outside. This method is ideal if you want to start slow-growing crops like tomatoes and peppers from seed.
- **Direct Sowing:** with this technique, the seeds are sown directly into the soil where you will grow them; no special equipment is needed, and no need to transplant seedlings.

Equipment:

Newbies often balk at growing from seed because they think setting themselves up with the equipment will be expensive. Here's the truth: you don't need a lot.

- **Seeds:** don't go overboard; you know what vegetables you want to grow, and you've worked out your companion plants. When you've worked out which ones you will grow from seed and which you will buy as seedlings, you know what seeds to buy.
- **Soil:** you can't use the soil from your garden for this; you'll need to buy some potting soil from your nursery. This contains the right mix of nutrients required for your seeds to germinate and grow.
- **Water:** don't use tap water if you can help it; it has too much chlorine. If that's your only option, place some into a jug and leave it for 24 hours at room temperature to dissipate the chlorine. Use clean rainwater or snowmelt if possible, and bring it to room temperature before using it.
- **Seed Trays:** these are for placing your pots of seeds in.
- **Pots:** for starting seeds, you can use 3-inch pots or root trainers (for certain seeds).

Different Types of Seed

We don't mean vegetable or herb types here. Different seed types grow in different ways, but when we break it down, there are two main categories: warm weather and cold hardy:

- **Warm Weather Seeds:** these will only germinate and grow in a warm environment. Too cold, and they'll do nothing, and even if they do, the seedlings won't survive. These are usually the best ones to start growing indoors and include peppers, tomatoes, eggplant, broccoli, basil, cosmos, zinnia, and marigolds, among many others.

- **Cold Hardy Seeds:** these seeds like the temperature cooler. If it's too warm, they won't germinate, or the seedlings will likely die. These are typically directly sown into the ground outdoors and include spinach, lettuce, radish, beans, beat, carrots, peas, sunflowers, and petunias.

Preparing to Grow Your Seeds

Before you start growing, you need to be prepared. Jump straight into it without following these next couple of steps, and you may not be successful:

- **Read the Packet:** this might sound daft, but you'd be surprised how many gardeners don't read the seed packets and wonder why they fail. Each packet tells you the growing requirements for the specific seed, including when to plant, whether to start them inside or outside, when to expect harvest, and so on.
- **Get Ready:** get all of your supplies ready before you start; this includes seed trays, pots, soil, etc. If your trays are older, you must clean and disinfect them before you use them in case they carry traces of disease or tiny pest eggs.

How to Plant Your Seeds

It doesn't matter if you are sowing in pots indoors or straight into the ground; the process is pretty much the same for both:

Step One: Prepare the Soil

If you are directly sowing into the ground, loosen off the top three or four inches of soil. Dig in some compost or worm castings and the organic fertilizer of your choice. If you are starting indoors, have a bag of high-quality seed compost.

Step Two: Work Out Your Spacing

This depends on what you are growing, and the spacing requirements will be written on the seed packet.

Step Three: Get Sowing

Again, a lot depends on the plant. Some seeds should be buried deeper in the soil than others. Make your hole and drop the seed in, or put it on top of the soil and press it down. The latter won't work with carrots and other fine, tiny seeds; these can just be sprinkled over the soil.

Step Four: Cover Them Up
Cover your seeds with soil and pat it down gently.

Step Five: Water Them
Spray water over indoor seeds. You want the soil moist but not washed out. Set your hose to a fine spray for outdoor sowings and spray it lightly over; do not disturb the seeds.

Tracking Your Plants
If you purchased a garden journal, use it now. If not, start a spreadsheet on your computer or just grab a notebook and pen. Write down the following:
- The seeds you just planted
- The planting date
- The gemination date
- Success rate – how many sprouted successfully

Make notes on which techniques you used and, as you track from start to finish, note what worked and what didn't, what you could do better, issues you faced, and so on.

Seedlings

Whether you grow your seedlings from seed or purchase them from a nursery or local grower, the process of handling and planting them is the same. You will need to transplant them from their current growing environment into another one.

What Is Transplanting?

Seedlings need to be transplanted into soil.

Transplanting refers to moving a plant from one environment to another, in this case, from one pot to another, whether purchased from the nursery or seedlings you grew yourself.

The biggest question you'll want to answer is when you transplant. That depends on the plant. Some crops must be plated before it gets too warm (like lettuce). Alternatively, warm-season crops, like peppers, eggplants, and tomatoes, shouldn't be planted until the weather has warmed up, as they don't like cool temperatures. Soil temperature is also an important factor. Check the forecasts; that way, you'll know what to expect.

Preparation:

If the weather ahead looks good, it's time to start getting your garden ready:

- **Prep the Soil:** over winter, your soil might have compacted, so loosen it up. Use a fork to dig it over and aerate it. Get rid of debris and weeds, and dig organic matter into it about a spade's depth down. This will help it drain properly but retain moisture, allowing roots to dig down.
- **Warm Up the Soil:** do whatever you can to warm the soil up; place black weed suppressant or plastic across it and leave it there a week or two before you plant.
- **Don't Walk on the Soil:** place boards down to walk on, or make a pathway with something else. If you walk on the soil, it will become compacted, and roots will struggle to make headway. And when you water, it will just run off.
- **Starve the Plants:** that's not as bad as it sounds! One week before you transplant your seedlings, reduce the amount of water you give them and stop fertilizing. This will help them adjust to life in the great outdoors.
- **Harden Them Off:** you can't just take a seedling from a warm, sheltered environment and plant it in a cold, outdoor spot. You need to transition it and give it a chance to get used to the change. If you don't, the plant will go into shock and may die. A week or so before transplanting, place the seedlings outdoors in a shady, wind-free area, but not too shady, as they will need to feel the sun. Do this for a few hours daily, gradually moving them out of the shade and into the sun and wind. This will help them get used to their new, permanent environment.

- **Moisten the Soil:** during hardening off, the soil must be kept moist, as outdoor air can rapidly deplete the moisture in the soil.

How to Transplant:

Try to choose an overcast but warm day and plan to transplant in the early morning. This will allow your plants to settle into their new home without full exposure to the hot sun.

1. Test the soil to see if it is too dry or wet to dig holes; it should be moist, not drowning in water. Add a lot of water to the soil twenty-four hours before planting. This will keep the soil workable when digging holes, and the roots will be watered immediately when planted.
2. Level the surface before you begin to dig and plant.
3. Place the plants (either in containers or removed from them) on the soil before digging to get an idea of the layout.
4. Start with the plants farthest from the edges of the area. Dig a hole larger than the soil and roots in the container to place the plant in.
5. Remove the plant from the container if you haven't already. Make sure you cover the soil side with your hand (do not damage the plant) and tap on the base of the pot – this will loosen the soil.
6. Pop the seedling into the prepared hole and fill in the hole with soil. Add a quarter-inch layer of soil over the top and tamp it down gently.
7. Settle the plant down by watering it generously. This will help acclimate the plant and fill any air holes left by digging and planting.
8. Give the plants time to fully acclimate (forty-eight hours) before you fertilize them. It is important to add fertilizer with phosphorous to help the roots take hold and for the plant to grow strong. Follow the directions on the label for adding the fertilizer.
9. If your planting bed is subject to hot weather, create ground cover to lock in the moisture by adding a layer of bark mulch on top of the soil.
10. Be mindful of the weather. If you have a bout of below-zero temperatures or a hail storm, protect your young plants by covering them to lock in the heat and keep out the elements. Remove the covering again when the weather returns to normal.

Never let the soil completely dry – you want to keep at least some moisture in the soil. Water it at ground level, which means don't hold the watering can or hose high above, as it can damage the plants – water daily until the plants are properly established.

Using Crop Covers

Plants that grow outdoors are vulnerable to the weather, the temperature, and many other problems that might come their way. This is why so many gardeners use crop covers to help protect their plants. Here are 10 reasons why you might want them in your garden:

1. Protect your young plants from digging garden pests like chipmunks, voles, and mice.
2. Speed up the germination process for direct-sowed seeds.
3. Shield tender plants from weather extremes and late frosts.
4. Shelter warm-weather crops from early frosts in the fall.
5. Keep the birds away.
6. Keep Mexican bean beetles, cabbage worms, hornworms, and other pests from laying eggs.
7. Cut down on the damage done by leaf-eaters like Colorado beetles, cucumber beetles, and squash bugs.
8. Stop deer from munching on your plants or brushing against your trees.
9. Keep cool-weather plants protected from the hot sun.
10. Keep your vegetables safe from rabbits, groundhogs, and squirrels.

Crop covers come in all kinds, each one offering its own type of protection. They are made from various materials of different sizes, some for single plants and others for several.

When Should You Use Crop Covers?

There is no specific time to use crop covers. You can use them as much or as little as you want, depending on what you're using them for. They can be used to protect from frost in early spring and late fall. If you want them to keep pests off your plants, use them throughout the growing season. And if you need to keep animals away, use them all year round. You get the picture.

The only other thing you need to know is not to wait until it's too late to use them. They are a preventative measure, not a solution for damage.

Here are the most common crop covers and their uses:

Row Cover

Row covers come in two types – plastic or fleece. Fleece is great for protecting tender plants from frost and is permeable, allowing moisture to seep through. Plastic creates a much warmer environment and is ideal for germination and bringing on seedlings.

You can extend your season with both row covers, but use plastic if your climate is colder. You will need to monitor them to ensure they don't overheat, and they will need to be secured down. Row covers allow you to get an early start on the growing season.

Best For: extending your season and protecting your crops.

Mulch

Mulch is incredibly versatile and is something every gardener should use. It locks in moisture and heat when the sun is out but also insulates on cold days. A mulch layer will also limit the growth of weeds. And, if you use organic mulch, it can help condition the soil when it breaks down.

The important thing about mulch is not to use too much – no more than a 4-inch layer around a plant. If you use too much, the plants can suffocate. Mulch can be anything organic that can be shredded and broken down to provide a coverage layer (paper, bark, straw, etc.).

Best For: perennial plants that need a little protection from cooler weather.

Cloche

These make great temporary covers to protect tender plants from unexpected frosts. You might think your frost season is over or not due to start for another month, but Mother Nature often has other ideas! It's best to be prepared.

The only downside is that cloches are not cheap, so using them on large areas of plants might not be cost-effective.

Best For: protecting small numbers of young plants.

Cold Frame

Incredibly sturdy, cold frames are usually wooden-framed with glass panes and a hinged glass lid to allow easy access to your plants. Winter gardeners commonly use these and can help you keep some food growing all year round, even when it's snowing.

The key to success is ensuring your plants are fully grown just as the weather becomes icy cold. Plant growth is much slower during winter, so you want them to be as mature as possible before putting them in the cold frame.

Best For: fall/winter gardening in colder climates.

Chapter 11: Watering and Caring for Your Plants

Growing a successful garden isn't just about popping plants in holes and hoping for the best. Whether you grow from seeds or starts, your plants need a certain level of care, which includes watering. When your plants are mature, their care moves to another level.

Caring for your plants is the key to their survival.
https://unsplash.com/photos/EdscD_R28bM?utm_source=unsplash&utm_medium=referral&utm_content=creditShareLink

Seeds and Seedlings

Typically, you will need to water seeds and seedlings every day or two, regardless of whether they are indoors or in the garden. Make sure to water evenly so all areas of the soil receive moisture. You do not want water pooling or dry areas.

That said, this will all depend on the soil type, temperature, other heat sources, and so on. When the weather is hotter and drier, or if you are using a polytunnel or greenhouse, you may even need to water daily, if not more. Make sure you use a moisture meter to check the soil moisture regularly.

Here's a quick guide on when your seeds and seedlings should be watered:

- **The Top ½-Inch (1 cm) of Soil Is Dry.** Most seeds are sown just beneath the soil surface, and seedling roots are short, so they need the soil to be moist around them. If the top cm of soil is dry, they won't fare well. Do not let it get to this stage; it could halt germination and stunt growth. Soil is dry when its color is lighter. If you don't have a moisture meter, poke your finger into the soil without disturbing the seed or seedling; if it's dry, water it.
- **The Tray or Pots Feel Light:** lift them daily and feel their weight. The lighter they are, the drier they are. With a little time, it won't take you long to learn when your plants need watering just by picking them up.
- **Check the Plants:** small seedlings are sensitive to changes in water; if not enough, they will start drooping. If you see seedlings like this, water them, but don't overdo it.

Direct-sown seeds and seedlings are a little easier to care for, and you have a bit more leeway. Pot-grown plants tend to dry out quicker than plants in the outdoor soil, and moisture is in shorter supply, whereas plants grown in the outdoor soil have access to a much deeper supply of water, which also prompts a plant to grow a deeper root system. They also benefit from dew in the mornings and rain showers.

As seedlings grow and age, they aren't quite so greedy for water. A week or 10 days after germination, you can get away with reducing the watering to every other day, and as they continue to grow, you can reduce it even further, so long as when you do give them water, it's a deep watering.

Under and Overwatering Your Seedlings

Obviously, if you don't provide enough water, your seedlings will dry out and can die, especially in hot climates. Older plants can be revived, even if they are very dry and a little wilted, but younger plants don't have the resilience to survive without water, even for a few days.

Another problem with under-watering occurs if you use peat moss in your growing soil. When peat moss dries out, it won't soak in the water; instead, it just runs off.

If your plants have dried out, water them as soon as possible; you might be lucky enough to catch them in time. If you have used peat moss and it is dry, soak it in a tray of water until it is rehydrated.

What if you overwater your seeds and seedlings? That's okay, isn't it? Well, no, it isn't. Many people make the mistake of flooding their plants with water when they've dried out, but this can lead to its own set of problems, including:

- **Root Rot:** when the soil is saturated with water, the roots can rot out.
- **Drowning:** yes, you can drown your plants because they can breathe. Water can fill the holes in the soil, stopping air from getting in, and the plants will drown.
- **Mold:** mold loves the damp, and it's fatal for young plants and seeds.
- **Damping Off:** this is a fungal disease that affects overwatered seedlings.
- **Insects:** some pests love the damp, and they'll attack young seedlings, which don't have the strength to survive.

What to Do If You Overwater Your Seedlings

If your seedlings are in trays, move them somewhere dry, airy, and sunny to help them dry out. If they are in the garden, you can only wait until the soil dries out and pray you don't get heavy rain.

The Proper Way to Water

Watering your plants isn't an exact science, but it must be somewhere near the mark for them to thrive. There are two ways to do your watering:

1. Bottom Watering

This uses the principle of capillary action. Water is taken from very wet to drier areas.

Place your pots in a shallow tray and fill the tray with water. Leave it be for a couple of hours; by then, the soil will have soaked up what it needs – you can check this with a moisture meter. If it is still dry, leave it for a bit longer. When the soil is moist enough, get rid of the remaining water. Keep an eye on things; your plants may be incredibly thirsty and go through the water quickly. If the tray dries out quickly, add more water.

This is one of the best ways to water seedlings as it is gentle, and the soil will only take in what it needs, which means no chance of overwatering.

2. Top Watering

Self-explanatory. This means you water from above. However, how you water from above depends on whether your seedlings are growing indoors in containers or outdoors in the ground. Those indoors are in lighter soil, which could be washed away if you are not careful, or you could break the seedling stems.

Here are the best top-watering methods for pot-grown seedlings indoors:

- **Mist:** use a spray bottle to mist water onto the seedlings, usually once a day or more. This will only water the soil surface and won't soak through the soil. You only need to do this until the seeds have germinated and started showing signs of growth; then, they'll need more water.
- **Lightly Sprinkle:** a watering can with a good hose on it offering a fine, light spray will work for this. Alternatively, make holes in the lid of a water or soda bottle and fill the bottle with water. If you use a watering can, use an indoor one –they're usually smaller, with thin spouts, and much gentler.

Watering your outdoor seedlings is as simple as using a hosepipe or watering can. However, the same principles apply; if they are young, don't squirt them with force. A gentle spray will do the trick. Alternatively, you could use a drop irrigation or soaker hose setup. Once the system is set up, simply attach your hose and leave it – the water will go deep into the ground around the roots.

Why Is Watering the Right Way So Important?

As mentioned earlier, it isn't an exact science, but it's also not about just tipping a load of water onto your plants. Understanding how plants use water takes time to learn, and other factors are also at play, including temperature, weather, soil texture, time of year, and time of day. You need to pay attention to all these factors because your plants will need different amounts of water at different times.

However, while you are getting to grips with your garden, there are some tips you can follow to help you:

1. **Water at the Roots:** water should be at soil level and applied until it has soaked the whole root ball. Don't forget, the root system on a plant can go a long way, so assume it is as wide as the actual plant and at least a foot down, if not more. That's where soaker hoses are best, as the water goes down through the earth to root level; 20 minutes and your plants will have all the water they need.

2. **Check the Soil:** don't water because you think you should; your plants might not need it. Use your fingertip to probe a couple of inches down into the soil; if it's dry, it needs water. Alternatively, use your moisture meter.

3. **Water Early:** the morning is the best time for watering, as the leaves can dry out during the day. If your plant leaves are constantly wet, there's a good chance that fungal disease will take hold. If early morning is not a good time for you, leave your watering until late evening, when the sun is not as strong.

4. **Water Slowly:** if you blast water at dry soil, it will run off and not sink in. Start slowly and build up; when the soil is moist on top, it will be better absorbed deeper down.

5. **Make It Count:** again, soaker hoses and drip irrigation systems are best. The water goes only where needed, and there's no waste. Also, early morning or late evening watering minimizes water loss through evaporation. For certain crops, the best companions are those that shade the ground, keeping moisture in.

6. **Not Too Much:** water isn't the only important thing to a plant; it also needs oxygen. Allow the soil to dry out a little between watering, especially with container plants. A deep watering once or twice a week is far better than daily watering.

7. **Don't Let Them Dry Out:** when the sun is at its highest and hottest, plants often wilt a little; this allows them to conserve some moisture, but you should see them perk up when the day cools down. If you don't see this, your plants have gotten too dry. This damages some of the tiny projections on their roots, and the energy it takes to regrow these should be going into plant growth.

That could result in stunted growth and poor harvest.
8. **Use Mulch:** using organic mulch around your plants can help retain moisture longer as it stops water from running off or evaporating. However, don't use too much mulch, as it can stop the water from getting to the roots.

Fertilizers

Plants need healthy soil to grow properly and produce fruit and flowers. All plants take nutrients from the soil. Some need more nutrients than are available. It's sort of a chain reaction:

- You feed the soil with nutrients
- Those nutrients feed the plant
- Plants feed us

Plants need three primary nutrients: nitrogen, phosphorus, and potassium. They cannot absorb nitrogen from the air, so they must get it from the soil, and if there isn't enough, fertilization is needed to boost it. Potassium exists in the ground but is usually far deeper than the roots go and isn't available to them. Phosphorus is also available, but only in certain rocks. The only way a plant can access this is if it is water soluble. That's why we need to give our plants a helping hand.

Organic fertilizers, such as compost and animal manure, are best, but you can also purchase fully organic fertilizers. Later, you'll learn how to make your own, but for now, refer to Chapter 9 for details on how to add fertilizers to your soil and plants.

Pruning and Deadheading

Experienced gardeners and newbies must follow routine maintenance to raise healthy plants, including pruning and deadheading. Not every plant will require this kind of care, but you must understand what to do when necessary. You might think pruning and deadheading are the same, but they are different methods and tend to be done at different times during the season.

Pruning

Pruning should be done regularly on some plants and is the practice of removing branches and foliage. You can use pruning to remove dead or diseased parts of the plant, trim shrubs into new shapes, or encourage growth.

Pruning promotes fresh growth, new flower buds, and a healthy plant. If you have old shrubs in your garden, pruning can give them a new lease of life. Look at it this way: just like you need to get your hair cut now and then to freshen and tidy it up, your plants need the same thing.

How to Prune

This is actually quite simple. Cut off what isn't needed using a sharp pair of secateurs or pruning shears. If you thin a plant's foliage, cut off up to a third of the stems. If you are pruning to cut a plant's growth back, don't cut off the stems, just the offending growth.

Annuals and perennials should be pruned once the first flowers have shown themselves and stop when the growing season ends. However, you must research because each plant is different and requires different pruning techniques and timings. Some plants have seasonal requirements and can only be pruned at certain times of the year.

Deadheading

Deadheading is an intuitive gardening process. Dead flower heads can still remove nutrients from the soil but won't grow. Removing them gives more food to the other flowers and flower heads. All you need to do is remove the dead heads.

You will also have a nicer garden when it is not decorated with dead flowers. When you remove dead flower heads, you might notice your garden bloom more and become more colorful now that the nutrients in the soil are not being wasted.

How to Deadhead

Simple. Just snip off any dead or faded blooms. Cut them off just above the first set of leaves below them. You can do this as often as you want or stick to doing it once a season; however, the less you do it, the fewer blooms you will have.

Not all plants need to be deadheaded. Typically, those that produce plenty of flowers, such as marigolds, roses, petunias, salvia, etc., benefit from deadheading. However, if they only have a single bloom, they won't thank you for cutting it off.

Chapter 12: Troubleshooting Common Companion Planting Issues

No garden is without problems and challenges, but understanding the problems and how to resolve them will help you keep your garden in good condition. This is especially true where companion planting is concerned, but thankfully, as it's a centuries-old technique, there's plenty of advice on potential problems. Here are the most common ones:

Insufficient Spacing

Many gardeners make the mistake of planting things too close together. You may not realize it then, but you'll soon see it when your plants mature to full size. You want your companion plants to do their job without crowding your crops.

How to Avoid:

Be sure to plan ahead. You do not want to allocate space based on the size of seedlings or seeds. Consider how big the plant will become, and space the plants out early – even if it looks like too much space when planting. If you are unsure how much space you need, always err on the side of caution and give more space than needed.

Competing for Water

This will depend on your soil type and its ability to hold water. If water is in short supply, hardier, deep-rooted plants will take the water and leave nothing for the others, a problem that also happens when plants are fighting for space.

Shallow-rooted crops struggle with water because the top of the soil dries out first, and they don't have taproots to help them get water. When companion planting, ensure deep-rooted plants can't take water from shallow-rooted ones.

How to Avoid:

Keep your soil moist. Use soaker hoses and mulch to keep the water consistent throughout.

Competing for Nutrients

Some plants need many nutrients, such as brassicas, tomatoes, cucumbers, squash, and peppers. The right nutrient levels allow them to produce a long season of fruit. However, they can steal all the nutrients in the soil, leaving nothing for other plants. If there are insufficient nutrients, the crops will suffer.

How to Avoid:

Pair your crops per their nutritional needs, i.e., don't plant companions that need the same nutrients as your main crops. Make sure to add plenty of organic fertilizer throughout the season to keep the levels up, particularly slow-release fertilizers.

Shading Your Plants

Companion planting offers fantastic benefits, but if one or more start overgrowing, they can shade the others out, and there go your benefits. Sunlight is the fuel required for plant growth and photosynthesis; if plants have to compete for light, it won't end well. Let's say you allow your cucumbers to grow along the ground rather than upright (not a good idea); taller plants will cut out their light and stop them from growing and fruiting properly. Likewise, tall tomatoes can crowd the light out from bush beans. However, some shade also benefits some plants, such as spinach and lettuce.

How to Avoid:

Almost all plants and flowers need sunlight to grow. When companion planting, ensure your taller plants are not providing too much shade. It can be tempting to over-plant a companion without realizing the damage it will do. For vine-like plants, provide support to raise them to get the sunlight they need. Study your garden before planting to see where the sun hits each day; this guides you better in terms of how to shade your plants.

Allelopathic Companions

Plants are a bit like people; they don't all get on together. Some plants stop others from growing, and this leads to certain disasters. Allelopathic plants produce chemicals from their roots that suppress growth in neighboring plants – basically, only the strongest will survive, and these plants are only interested in themselves.

How to Avoid:

Be careful with your companion planting and ensure no allelopathic plants exist. Here are some of the worst combinations:

- **Alliums or Mint with Asparagus:** Alliums and mint are both good at pest control because of their volatile oil production but can reduce the growth of companion plants
- **Onions and Beans:** They do not work well together and will limit growth, especially from seed
- **Sunflowers and Potatoes:** Sunflowers limit the growth of potatoes with the chemicals they release; sunflowers also produce too much shade.
- **Fennel and Almost Everything:** The compounds released into the soil by fennel stunt the growth of almost everything around them

Different Soil Requirements

Changing the soil composition from one spot to another is almost impossible, and not all plants thrive in the same conditions. Some plants prefer alkaline soil, while others prefer acidic soil. Some plants can thrive even when the soil is changed, starting as a great companion before becoming an enemy. Plant these near one another, and one of your crops won't grow properly.

How to Avoid:

Understand your plants' soil and pH requirements, and only plant those companions that work with your crops, not against them.

Bad Timing

Timing is important with companion planting. The right companions will complement one another perfectly throughout the entire growing season. For example, if you plant tomatoes, fill the rest of the bed with lettuce or radishes. They are quick plants that will be fully harvested when the tomatoes are fully grown.

Some plants must be established before their benefits as a companion become apparent. For example, if you plant corn and cucumbers together and want the cucumber to use the corn as a trellis, the corn must be at least a foot or two tall before you plant the cucumbers.

Lastly, flowering must be considered. The flowers offer the best companion planting benefits, so while the foliage on many plants does a good job at repelling pests, their flowers bring beneficial insects and predators in.

How to Avoid:

Look at the DTM (Days To Maturity) of every plant. Maturity is usually the time taken to reach the first harvest – some plants will continue fruiting all season. Synchronize your planting times so everything reaps the benefits, and also experiment; try staggering plantings to see what works.

Aggressive Companions

Some plants are overly aggressive and really shouldn't be planted beside vegetables:

- Bamboo
- Bee Balm
- Blackberries
- Clover
- Jerusalem Artichokes
- Mint
- Morning Glory
- Rosemary

- Thyme

These plants can be stunning but grow so fast, choking everything else in their path. Some will spread fast underground – bamboo, rhubarb, mint, etc., and pop up all over the garden, bypassing any barriers you might have put up.

How to Avoid:

Don't plant these anywhere near your vegetable beds. Plants like rosemary, mint, rhubarb, and thyme can be planted in containers. That way, you get their companion plant benefits without the downsides they come with.

Haphazard Plantings

You might think a haphazard, messy planting shape is fun, but rows are there for a reason. They are visually appealing and make it easier to tend to the plants. Watering (as does weeding) becomes hassle-free, and you can easily see which plants are growing and which are not.

How to Avoid:

Keep the plants in rows. This is easy to do with some forward planning and will make life much easier in the growing stage.

Different Maintenance Requirements

You should already have discovered that the real secret to successful companion planting is to plant like with like. That means planting those with similar requirements together. If you plant companions with different requirements, things will almost certainly go wrong. Some plants require you to add additional soil during the growing season, and if you plant them with low-growing plants, the low-growing plants will be covered.

How to Avoid:

Be sure you know what the plant needs before planting it. A little research can go a long way toward successful plants and reduced problems.

Wasted Space

If you only have a small garden space, companion planting is a fantastic way to make the most of your space and increase your yields. However, not understanding how to time your crops, when and what to trellis, and how to space everything could lead to a lot of wasted space.

How to Avoid:

Map all the bare spaces in your garden and determine how to fill them. For example, when you plant young tomatoes 1 to 2 feet apart, you have a lot of empty space. That could be filled with fast-growers, like lettuces, spinach, or radishes. By the time the tomatoes reach maturity, those crops can be harvested. If you put another slow-grower, such as peppers, into a bed, interplant with basil or scallions to fill in the empty gaps.

Attracting the Same/Similar Pests

When two plants attract the same kind of pests, planting them together is sure to end in disaster. With a choice of crops to attack in one place, pests are more likely to settle in, breed, and destroy your garden. This happens because the crops look and/or smell the same. For example, two brassica family members, cabbage and kale, attract aphids in large numbers. Plant them together, and you have a huge problem on your hands. By interplanting with onions, sweet alyssum, or calendula, you can stop those pests from jumping between your plants; even better, plant some nasturtiums a distance away. These are trap crops that attract the aphids away from the crops.

How to Avoid:

Select your companion plants carefully in terms of the pests they attract. Do not plant the same family of plants too close to each other, such as:

- **Brassica Family:** broccoli, Brussels sprouts, radish, cauliflower, kale, mustard
- **Cucurbitaceae Family:** squash, cucumbers, melons, zucchini
- **Solanaceae Family:** potatoes, eggplant, peppers, tomatoes
- **Amaranthaceae or Chenopodiaceae Family:** chard, beets, spinach, quinoa

If you have to plant them close together, plant companion plants that are excellent at deterring pests, or your garden will be overrun.

Too Many Companions

Companion planting is fun, but going overboard when you are a newbie is easy. When you add too many plants to one bed, you create more problems than you solve. The garden becomes overgrown, and you will struggle to tend to any of your crops; not only that, but you also won't be able to see which companions are working.

How to Avoid:

Having a diverse garden is wonderful, but aim for simplicity when you first start. Try a couple of combinations per garden bed to learn what works together and what doesn't. As you gain experience, you can experiment a little more.

Not Using Groundcover

Companion planting isn't just about keeping pests away; certain plants act as ground cover to keep moisture in and weeds down and as a habitat for beneficial insects. These plants may not always look colorful or smell amazing, but they do a fantastic job in the garden.

Ground-cover plants can:

- Limit weed growth
- Provide ground cover to keep heat from the soil
- Elevate other plants to reduce rotting when fruits are in contact with the soil
- Replace a layer of mulch
- Lock in moisture in the soil for long periods
- Enhance the soil with root growth

How to Avoid:

If you do not have ground cover, planting a low-growing plant can add that to your growing area. Micro-clover is great and will also add nutrients to the soil; creeping thyme is another hardy companion. Pair them with taller plants that provide shade where necessary, and half your work is done by the plants.

Companion planting is not complicated, but you do need to do some planning to keep the mistakes to a minimum. Before you decide on your companion plants, ask yourself these questions:

- Have I given each plant enough space to reach full size?
- Will these plants fight for nutrients and water? Are the fertilization needs similar?
- Will one plant grow bigger than the other and shade it? If yes, does the smaller plant benefit from a little shade? Does it need a

lot or a little?
- What does the plant do for pests? Will the same pests attack both plants?
- Is one of the plants allelopathic? Will it attack the other one?

If you don't create a solid plan, you might end up with a disaster zone in your garden. However, if you do try a combination that doesn't work, don't worry; we've all done it! If it causes too many problems, pull up the companion plant and replant it elsewhere in the garden.

Companion planting takes time, but keeping a notebook can help; note how your plants work together or cause problems so you know what to do next time.

We've finally got there – it's harvest time!

Chapter 13: Harvesting Your Organic Companion Planting Garden

You've put in all the hard work, and now it's time to reap the rewards - that succulent harvest just waiting to be picked. But how do you know when it's the right time?

That is one of the most common questions gardeners ask, and that's because most inexperienced gardeners have a preconceived notion of how their fruit and veggies will look - just like it does in the grocery store, right?

We also pick too soon, impatient to get our hands on what we've grown, or we go the opposite way and leave it too late.

First, here are some tips to help you with your harvest:

1. Get in Your Garden Daily

When your harvest starts ripening, it can all happen simultaneously; that's why you should be out there every day. If you don't wander your plants regularly, you will miss early ripening produce and leave it to rot. This attracts pests and diseases, which can soon wipe out your garden.

You don't want any of this, so check your plants every day. When you spot ripe fruits and veggies, pick them. This does two things - it gives you delicious food to eat, and in some cases, it can encourage a plant to continue fruiting - tomatoes and cucumbers are two examples.

2. Pick Small

How often have you looked at your zucchini and thought, *I'll just let it get a bit bigger?* Before you know it, it's huge, which is not good. If you let your veggies grow too big, they lose flavor. Smaller vegetables taste better, are more tender, and don't have too many seeds.

That said, if you do find a huge zucchini or a tomato that looks like it ate an entire box of growth hormones, don't chuck them; you can still use them in your cooking.

Pick small and enjoy more flavor while encouraging your plants to continue producing.

3. Gently Does It

You can involve the kids in harvesting, but they must be gentle – it doesn't take much to bruise fruits and vegetables. They must be picked off the plant gently and placed in the harvest container even more gently!

It's not because bruised produce doesn't look nice; it can hasten rotting, leading to a much shorter lifespan of your harvest. If you do bruise any, you need to use them immediately.

4. Make Sure Your Harvest Containers Are Large Enough

Ensuring you place your harvest into large enough baskets will lessen the chance of bruising, which means you can bring in a bit more each time. 5-gallon buckets are ideal for beans, and large baskets (bushel-type) are best for larger plants, like squash, eggplant, cucumbers, etc.

You could use a clothes basket if you have nothing else or even a wheelbarrow, but be careful with softer fruits.

5. Look Where You Are Walking

Make sure you have clear pathways between your plants.
https://unsplash.com/photos/1_yycyoMT6g?utm_source=unsplash&utm_medium=referral&utm_content=creditShareLink

This is important, especially if your garden is well-grown. Create clear pathways between your paths, or be careful where you walk. You may step on smaller plants or lower fruits and vegetables. Not only will this damage the crops, but it can also open up the doorway to pests and diseases, quickly wiping out your harvest. If you do accidentally step on a vegetable or fruit, pick it up and dispose of it immediately.

6. Keep Track

Keeping up with everything is difficult when you have a busy garden full of different plants. You need to know what crops you planted, each variety, the time to harvest, and what you should look for in a harvest.

Keep a garden journal – you can buy specially designed ones or use a notebook – and write down everything. When you know all this information, you know when you need to start checking for harvest. You can track the plant growth in your planner and note down the start of the harvest season so you are ready to harvest – but also know not to be away from your garden during this period. It can be tempting to harvest too early, and it is not uncommon to harvest too late, but by noting down when you should harvest based on the plat type, you can avoid these common problems.

7. Monitor for Disease

When the seeds are sprouting and right before harvest – these are two critical times to check for any possible disease. Check for misshapen leaves, discoloration, or dead parts. Check your plants regularly; it can help you pick up on early signs of trouble and take steps to correct it. If you do find disease or pests, look at your companion plants and decide if you could plant something different next time to stop the same thing from happening again.

8. Be Realistic

Don't base your expectations of your plants on what you see in stores or on seed packets. For example, the heads on home-grown broccoli don't tend to be as large as those from the stores. If you have unrealistic expectations, you won't know when to harvest; you'll be looking for something that doesn't exist. Leave them too long, and they will soon start to rot.

Knowing what to expect from your harvest is the easiest way to know when to pick it.

9. Quickly Harvest Stems

The stems that are not producing fruit or flowers should be removed as quickly as possible to allow more nutrients to go where needed. Leafy vegetables and herbs should be harvested early to lock in the flavor – leave them too long, and the flavor will deteriorate as they use more nutrients.

10. Allow the Fruits to Hang

Those plants that produce fruit, such as apples, tomatoes, peppers, etc., shouldn't be picked too early, or they won't be fully ripe. Leave them to ripen fully before you harvest them; this comes back to knowing your varieties and what to look for.

Let's look at some of the most popular crops so you get an idea of the best time to harvest them.

Popular Plants – Harvest Times and Methods

Most plants can be harvested without the need for special tools, just a pair of gloves and a basket. However, in some cases, you can use pruners or a small knife to help you out. Here are some of the most popular plants you are likely to grow and tips for harvesting.

Herbs

Learn what your herbs should look like when they are ready for harvest. Then, you should cut them back as often as possible and store them in the refrigerator on dry paper towels to soak up the moisture. Alternatively, hang them somewhere cool and dry where they can dry out before you store them for harvest. You can also chop herbs and place them in ice molds with oil. They can then be used later individually.

Tomatoes

Tomatoes come in various colors and sizes depending on the type. A tip here is to look at the seed packet if you grew your own, or the nursery label may have a picture of a mature plant on it. The fruit should be firm but gives a little when gently squeezed.

Ripe tomatoes should come off the stem easily. Pull on them gently; if they come off, they are ready.

Peppers

Peppers will also change color as they ripen, but many varieties can be picked at any color. Expect to see green, yellow, orange, and red. Peppers will sweeten over time, so the longer they grow, the sweeter they taste. Be sure to check the harvest time of the peppers you are growing. You can grow them for too long if you are not careful.

When harvesting, cut the peppers from the plant rather than pulling them off. Hold the stem and twist the pepper if you don't have a knife handy.

Lettuce

Lettuces can generally be harvested when the leaves are about 4 inches long, depending on the variety. Try to pick them while it is still cool outdoors; in extreme heat, lettuce can start to produce seeds, giving the leaves a bitter taste.

Leaf lettuces should be harvested from the outside in, while head lettuces (iceberg, for example) should be sliced off at the stem. Most lettuces are known as cut-and-come-again, which means they will continue to produce.

Green Beans

Beans are another crop that continues to produce when picked. You can enjoy a good harvest all season with just a few bean plants. When you see blossoms on the plants, start checking; pick the pods young for a sweeter bean. Don't leave green beans for too long, or they will no longer be tender and soft – check the seed packets for harvest times.

Do not pull on them too hard; you might end up pulling the whole plant out. Use scissors or secateurs to snip them off. Do not harvest beans in the morning as they will likely still be wet and dewy, which can spread disease.

Green beans are the perfect example of natural fertilizer. Once the plants are finished providing beans, chop them down and leave them on the ground. Let them start to die off and dig them into the ground, roots and all; this is the perfect way to boost your soil with nitrogen.

Peas

This is a trial-and-error crop in terms of harvesting. Check the peas regularly by opening a pod and tasting the peas. If the peas are the size you want, continue harvesting; if not, leave them a bit longer. Again, once the harvest is done, dig the pea plants back into the ground for a boost of nitrogen.

Melons

Seriously, testing a melon for ripeness is as simple as thumping it. If it sounds hollow, it's ready. If you don't want to do that, have a good sniff;

most melons give off a sweet aroma when they are ripe.

Harvesting is as simple as snipping the fruit off the vine.

Watermelon

Check the part touching the ground to see if your watermelons are ripe. The melons should be green and stripy and have a yellowish spot on them where they lay on the ground. The fruit is not ripe if that spot is white or light brown. Again, simply snip them off the vine.

Cucumbers

Knowing the variety you planted can help you determine the size of the cucumbers. When they get to that size, harvest them. Leave them too long, and they will produce a lot of seeds inside and be bitter. Do check your plants thoroughly; they are quite leafy, and you might miss one or two that grow into massive fruits.

Harvesting is done by tugging and twisting gently to remove them. However, you can also use secateurs or scissors to cut them off.

Corn

When your corn starts to form ears, give them a gentle squeeze. A husk covers the corn cob, but you can feel the corn beneath. When the strands of husk begin to dry, check a corn kernel if you have access to it. Squeeze the kernel between thumb and forefinger - if white sap emerges, your corn is ready.

Corn husks are easy to remove from the stem when ready.

Root Vegetables

Keep track of all your root vegetables so you know when to harvest, as the harvest time is tricker to determine by examining the plant alone. When ready to harvest, gently pull one plant to check the size of the vegetables. With carrots and beets, you can thin them out earlier in the season. This means pulling every plant out - you should have fully formed baby carrots or beets, which are delicious. This leaves the rest of the harvest room to grow - two crops for the price of one.

Garlic

Check the tops of the garlic to determine whether it is time to harvest. The tops will turn from green to brown when the garlic is ready. Don't do any prep or cleaning other than hanging the garlic to dry once it is removed from the ground.

Eggplant

Eggplants can be quite bitter if you let them get too big. Instead, harvest them when small and purple with a nice sheen; they should also be firm to the touch. Do not pull eggplants off the plant; you will destroy the entire plant. Instead, snip them off and allow your plant to put its energy into producing more fruit.

Onions

Onions take a long time to mature; you usually won't gather in your harvest until the end of the growing season. Like the garlic, look at the tops; you can harvest your onions when they have died off. However, if you see an onion with a long, thick stem in the middle and a flower head, cut it out and harvest it immediately – left in the ground, it will go tough and may rot.

Harvesting is simple: pull them from the ground. If you are storing them over the winter, they must be cured. Lay them in a single layer with space around each one and let them dry. If your weather is warm and breezy, they can be laid on top of the earth to dry.

Potatoes

Potatoes are ready for harvest when the leaves have turned yellow and died back. To see what's grown, root around in the earth at the base of the plant; alternatively, just let the plants die back and dig up the spuds.

Harvesting depends on your growth method. If you grow them in mounds or trenches, carefully remove the plant and dig around in the soil for the potatoes. If you use a fork, be careful not to stab any. If you damage any potatoes, these must be used quickly and cannot be stored, or they will rot – one rotten potato can cause all the others in storage to rot. If you grow your plants in potato bags, tip them onto a tarp and sift through the soil.

Carrots

Carrots will be ready to harvest in around 7-8 weeks, depending on your climate – smaller varieties of carrots will not take as long. When you are ready to harvest, simply pull them from the ground. Carrots are robust, and you can leave them in the ground until you are ready to eat them instead of pulling and storing them – just be careful to watch for frost.

However, unless you are growing them in a polytunnel or your climate is quite warm, you cannot leave them in the ground over the winter; they will freeze into the soil.

Kale

Kale is a simple plant to harvest.
https://unsplash.com/photos/M8bpp4qQZGg

Kale is a simple plant to harvest; just wait for the leaves to be large enough. This will normally be around 10 inches long, although you can pick them slightly smaller if you prefer.

Harvest the outside leaves first; kale plants will grow new leaves and produce large quantities throughout the season. These are usually hardy and can survive winters on the ground.

Summer Squash

Summer squash matures quite quickly, so long as they are pollinated successfully. The fruits are quick to grow and can be harvested at the size you prefer; the bigger they are, the more seeds they will have. Most summer squash varieties take about two months to mature.

You don't want to pull squash, or you can damage the flower; use a blade to cut them from the stem (always use a clean knife to limit the spread of plant diseases). If you do damage the stems, they might not produce any more squash.

Snap Peas

This variety of peas can be harvested in around seven weeks. Harvest them early before they become tougher and more stringy. Check these daily around harvest time.

Do not tug the pods off, as you will damage the plant. They should snap off easily.

Brussels Sprouts

Sprouts are usually ready to harvest when they are around one to two inches in diameter. These are a slow-growing crop, though, so be patient. You usually won't harvest until the end of the season.

Pull sprouts off as needed or harvest the entire plant to keep the pests off. They can be blanched, frozen, or stored in the refrigerator for up to two weeks.

Beets

Beets may be harvested as babies or mature. The average maturity time depends on the variety you grew, so check the seed packet. If you leave them too long, they will become tough and won't taste very nice. Like carrots, they cannot be allowed to freeze in the soil, but a touch of frost won't hurt them.

Harvesting is no more difficult than pulling them from the ground. You can also eat the leaves; add them to salads or stir-fry them with a little salt and garlic.

Spinach

Most spinach varieties will bolt very quickly, so keep an eye on them and harvest the leaves as often as possible – baby spinach leaves are lovely.

Pull the plant and use the leaves if it looks like it will bolt. Otherwise, just pick off leaves as you want them, and they will regrow.

Keep a constant eye on your garden; you will soon learn what's ready to harvest and what isn't. It also means you won't lose plants to bolting or end up with bitter fruits because you left them too long.

Bonus: Organic Fertilizer Recipes

Organic fertilizers are available in almost every nursery or garden center, but why buy them when you can easily make your own? Our final chapter provides you with some easy organic fertilizers to make using what you would normally just throw away.

Grass Clipping Tea

Fresh grass clippings are full of nitrogen, and you shouldn't add too much to your compost pile. However, you can use them as mulch - don't put them too close to your plants, as the grass is acidic and can burn them. Layer the clippings no more than two inches deep; any more, and they will flatten into a wet mess that won't allow oxygen through and can cause your plants to go moldy. You can also make a nitrogen tea to feed your plants:

1. Take a five-gallon bucket, and one-third fill it with fresh clippings. Fill the rest of the bucket with clean water.
2. Leave it for two weeks, stirring now and then.
3. Strain the grass from the liquid and mix it one part grass tea to five parts water - it should be like a weak tea. This should be applied at the soil level, not on the leaves.

Manure Tea

If you can get hold of fresh livestock manure, you can make a tea your plants will love:

1. Fill one-third of a container with manure and two-thirds water.

2. Leave it for three days, stirring now and then
3. Strain the tea and throw the manure on the compost heap
4. Dilute the liquid with water until it turns a pale brown-yellow, see-through liquid.
5. Again, this should be applied at the soil level, not on the leaves, especially on spinach, lettuce, and brassicas.

Dandelion Tea

Dandelions are full of potassium that plants need for photosynthesis, and you can use the entire plant to make tea:

1. Harvest your dandelions – the tops of the whole plant; it's up to you. Do NOT use any that have been sprayed with herbicide.
2. Put a good bunch of dandelions into a five-gallon bucket and top it up with water.
3. Put a lid on and leave it for three to four weeks, stirring now and then. As the dandelions break down, there may be a smell, and the water will go black.
4. Strain it and discard the dandelions onto your compost heap.
5. Dilute the tea to a light color and apply at the soil level. This will encourage the plant to flower and produce fruit.

A word of advice on dandelions: avoid spraying them with chemicals, and don't pick them or mow them down too early in the season. They are often the first food source for pollinators like bees; if you kill them off, the bees can't feed.

Banana Peel Tea

Another great way to feed your plants potassium is to make tea from banana peels.

1. Gather enough banana peels to fill a container. If you don't eat that many bananas, chop the peels of what you do eat and freeze them. When you have enough, you can dump them in the container.
2. Fill up the container with water and leave them for one or two weeks
3. Give it a good stir, then strain it; the remaining peels can go on your compost heap, and the tea can be diluted one part to five

parts water until it is a light color.
4. Feed at the soil level until needed, or use it as a foliar spray to deter aphids.

Crushed Eggshells

Eggshells are full of calcium and can help raise soil pH. However, if you just throw them in your garden or the compost heap, they won't break down for years, meaning calcium isn't readily available. The quickest way to solve this is to crush or grind them:

1. Save your eggshells and spread them on a baking pan. When the oven has been on, place the pan in and dry the eggshells for a few minutes until brittle. This will also kill off salmonella bacteria.
2. Grind the eggshells to a powder consistency or crush them down into a fine consistency.
3. Apply as a top dressing around plants that need calcium, dig them into the soil, or add them to one of the fertilizer teas.

Coffee Grounds

Coffee grounds usually get tossed in the trash but make a great fertilizer packed with magnesium, nitrogen, and potassium.

1. Spread your coffee grounds on a tray and let them dry out.
2. Once dried, you can sprinkle them sparingly around your plants.
3. You can also add these to one of the teas mentioned above.

These are perfect for plants that love acid, i.e., rhododendrons, azaleas, blueberries, and roses.

Epsom Salts

Most people have a box of these lying around; if not, they are readily available. They are full of two secondary nutrients – sulfur and magnesium.

1. Add one tablespoon of Epsom salt to one gallon of water and stir it thoroughly to dissolve the salt.
2. Use this to water your plants once a month throughout the season, especially tomatoes, potatoes, peppers, and roses.

Vinegar Fertilizer

Vinegar adds acidity to the soil. If you have plants that need acidity-rich soil, white vinegar is one of the best things to add to your fertilizer. Vinegar is great for house plants as it does not harm any children or pets.

1. Add one cup of white vinegar to a gallon of water and stir it in
2. Water your plants with this once every three months

Never use undiluted vinegar on your plants, as it will kill them.

Compost Pile

Making a compost heap is one of the best ways to feed and fertilize your soil. You can do this directly on the ground, build or buy a proper compost bin, or just use a bin. Throw in all your vegetable and fruit scraps, some grass clippings, and any other compostable material – make sure you add cardboard, newspaper, or shredded paper, as this balances the compost and helps it turn quicker. Add a little water now and then and turn it to speed up the composting. You can also purchase rotating composters – add the material, shut it, and turn the handle.

Although it takes a while to make compost, when it's ready, it will feed your soil with microorganisms and nutrients that help feed your plants during the next season.

Mix and Match

Commercial fertilizers are usually a combination of nutrients, and you can emulate this in your own home:

- When you make your grass-clipping tea, add a tablespoon of Epsom salts and some banana peel to the container
- Combine your dandelion and grass clipping teas, and add a healthy dose of crushed eggshells

Get creative; have fun and make lots of notes so you know what works for next year

Conclusion

Whether you knew nothing about companion planting and gardening or were already experienced, I hope you now have more knowledge to put to good use.

Companion planting is an important part of gardening. It's not about making your garden look pretty; although done correctly, it can have a stunning visual effect. It's about gardening organically, about using plants to keep pests at bay and control weeds without using chemicals. It's about helping the soil structure, feeding it with nutrients, and helping other plants to thrive and produce a bountiful, healthy harvest.

This easy-to-read guide has provided you with information on everything you need to know, including:

- What companion planting is and how it all started
- How to plan your garden
- What plants make good companions, and what don't
- How plants help each other
- The difference between starting from seed or buying starts
- How to plant your garden
- How to care for your garden
- How to harvest your bounty
- How to make and use organic fertilizers

If you are a beginner, this book has hopefully awakened your passion to get out in the fresh air, get your hands dirty, and produce a gorgeous, healthy, organic garden.

Part 2: Cacti and Succulents

An In-Depth Guide to Maximizing Yield, Quality, and Beauty, Along with the Best Companion Plants for Beginners

Introduction

Cacti and succulents have been around for generations, but did you know it has been only a few hundred years since humans first began the domestic cultivation of these prickly plants? In the grand scheme of things, cactus farming is still in its infancy. The progress of succulent cultivation has been anything but consistent throughout the ages. It reached a new high on the graph only recently. But that doesn't mean humans don't know much about these thick-leaved plants.

In this book, you will explore everything that humankind has managed to learn about cacti and succulents, what they are, what makes them unique, their scientific names, their uses in ecology and at home, and an explanation of why you should start cultivating these water-retaining delights if you haven't already begun your own garden yet.

Once you are pumped up and ready to begin your journey into the growth of cacti and succulents, you will be introduced to the various species that originated in the Wild West, their natural habitats, and how you can domesticate them. After all, they live in some of the harshest known conditions. Will adapting to the more tolerable atmosphere of your house cause any problems for the plants?

Then, you will dive into the deep recesses of the planting process of cacti and succulents, how they can be grown from different base mediums, their watering cycle, their need for fertilizers, and overall maintenance. Cultivating and maintaining cacti and succulents is much easier than growing and nurturing other types of plants, and you can start reaping the benefits of these thorny plants in no time.

As soon as you have your first succulent fully grown and blooming in all its glory, you will be filled with the urge to cultivate another one. You don't need to follow the same lengthy procedure to develop your second cactus. Later in the book, you will learn the art of propagation and find an in-depth guide to cloning.

By the time you have finished reading, you will not only be able to tackle any challenges you may face post-cultivation but also have a short list of some of the finest companion cacti and succulents at your disposal.

Despite the complicated nature of these bulbous living plants and their biological importance, you will find the information in this book to be accurate and amazingly easy to grasp. It is perfect for beginners, but a seasoned horticulturist will also find a few illuminating facts and techniques here. Get ready to enjoy a wealth of knowledge on cacti and succulents, accompanied by some spectacularly vivid illustrations, which will lead you into this fascinating world.

Chapter 1: The Fascinating World of Cacti and Succulents

Cacti and succulents aren't worlds apart, but they aren't quite the same either. Every cactus is a succulent, but every succulent may not be a cactus. It's like saying every human is an animal, but not all animals are human beings. "Cactus" is the name given to a botanical family, whereas a succulent is a group of plants of which cacti are a part.

You may know what cacti look like. After all, they have been described in various novels and shown in a number of Hollywood movies. They are thorny green bulges that sprout from the ground, devoid of any leaves, typically found in a desert.

Quite bland on the surface, but did you know that they can grow some of the most beautiful flowers? The colors of these flowers are their most remarkable feature. From a brilliant pink to sapphire blue, a sea of full-grown cacti is a vibrant kaleidoscope of scintillating colors that will please even the least artistic eye.

Cacti store water in their thorny spines. These spines act like leaves (they actually aren't), and they perform photosynthesis whenever required. Succulents that are not cacti, on the other hand, usually have leaves that can store water and other nutrients.

That said, there are succulents with spines that aren't cacti. In essence, the absence of leaves is one of the few differentiating features between cacti and succulents.

Cacti are native to the barren lands of the New World, or as it is called today, the Americas. In other parts of the globe, they are domestically cultivated rather than naturally grown. That's because they are some of the easiest plants to nurture.

A sea of full-grown cacti is a vibrant kaleidoscope of scintillating colors that will please even the least artistic eye.
https://pixabay.com/photos/cactus-flower-botanical-desert-2721269/

A Brief History of Cacti and Succulents

These days, most people connect cactus with a spiny plant found in arid regions. But the name was originally used in Greek as káktos long before the cactus was discovered in the Americas. Káktos is a spiny sow thistle plant commonly found throughout Europe, northern parts of Africa, and the western regions of Asia. Intrinsically, they are not associated with cactus in any way and have entirely different taxonomies. They only share similar names and spines.

The cactus is a part of the Cactaceae family of plants, a family with an ancient history going back millions of years. However, the cactus you know and love today is relatively new because its fossils haven't yet been found.

Every species of cactus, except the Rhipsalis baccifera, which is found in Africa and Sri Lanka, naturally grows only in the Americas. This fact has given rise to an interesting theory. In an age long past, an era when dinosaurs roamed the Earth, there were no continents, just a single extensive landmass, a supercontinent you could call it. This landmass did not have any cacti, just regular ancient plants, many of which are now extinct.

Then, some 200 million years ago came the breaking up of the world. A huge chunk floated off far to the west, and it was there that first traces of cacti poked and scratched their way to life. This huge chunk is now called the American continent. And since cacti originated after this break, their natural growth does not occur anywhere else in the world.

The modern-day cactus can be traced back to the Aztec civilization (early 15th century), thanks to their clear and accurate drawings. The most prominent pictorial representation (also found on the coat of arms of Mexico) shows a majestic eagle perched atop a classic three-stemmed cactus. When Christopher Columbus discovered America, he carried the first cactus plant to Europe on his return journey.

Succulents, the plants, are so called because of their amazing ability to retain water in their stems and leaves (*sucus* in Latin means "sap"). The reputable Portuguese explorer Vasco da Gama is credited with the discovery of succulents in the southwestern part of Africa. It probably happened around the same time that Columbus discovered America.

Succulents, like cacti, first grew in the semi-arid and arid regions of the world, but unlike cacti, they are naturally found not only in America but also in Africa, Europe, and several areas of Asia.

Since succulents and cacti were present hundreds, and even thousands, of years ago, they were once (in many instances, they still are) an integral part of many cultures.

Cultural Significance

The roots of cactus run deep into the culture of Mexico. After all, the first sightings of the plant in modern civilization were experienced by the Aztecs. As you may know, the Aztec Empire was spread throughout the central part of what is now called Mexico. In fact, the most commonly found species of cactus in Mexico, the nopal cactus, was at the core of the founding of the Aztec civilization. Here's how the story (or myth, depending on your belief) goes.

Aztecs and Mexico

The Aztecs were native to a place called Aztlan, rumored to be somewhere in the southern United States. It was when their god of sun and war, Huitzilopochtli, ordered them to leave Aztlan and find a new home that they set out southward. The god had given them a landmark – see an eagle devouring a serpent atop a prickly pear cactus (nopal) – and that was where they should begin their civilization. They found that landmark in what is now called Mexico City.

Today, apart from being a part of Mexico's national emblem, the cactus also has a significant impact on the culture of its people. Many consider it to be the "plant of life" since it can survive without water for, apparently, several years (in reality, it can survive for a month or two). It is a symbol of hope for many because, after enduring several obstacles, the Aztecs were finally able to find their home.

Native Americans

Mexico's northern counterparts, the Native Americans, also have a high regard for cactus. Their legends go something like this: The cactus was first born after a woman buried herself alive in the ground. It is believed that she rose from the soil as a magnificent cactus with arms stretched toward the heavens. A beautiful circlet of flowers adorned her head every spring, followed by a crop of sumptuous fruit called *bahidaj*.

The Native Americans hold the cactus sacred for its ability to sustain its own life and that of others through its fruit. Its mythology and the beauty of its flowers also bring spiritual comfort to their lives. The Native American affinity for cactus is artistically depicted by the 20th-century painter Ted DeGrazia, in many of his works.

Japanese

Though cacti don't occur naturally in Japan, they are grown throughout the country. The *hanakotoba*, or the art of learning the language of flowers, depicts the flowers of the cactus as a symbol of sexual lust. When a man presents a cactus as a gift to a woman, it implies that he is romantically interested in her.

Chinese

In the Chinese practice of Feng Shui, cacti are considered to bring bad luck. However, if placed on the porch or in the backyard, cacti are believed to repel bad luck from outside, keeping your house free of negative energy.

Succulents

Each succulent has different meanings in many cultures. Aloe Vera was a symbol of beauty for the ancient Egyptians, and it was used to heal wounds during the reign of Alexander the Great. The jade plant is known as the "fortune plant" or "money tree" in various cultures. It is believed to bring positive energy into the house.

Dracaena trifasciata (snake plant) is associated with storms and is generally believed to improve the air quality in its surroundings. Echeveria elegans (Mexican snowball) is native to the northeastern region of Mexico, and it usually symbolizes strength and endurance.

All these incredible cultural and mythological claims about the power of cacti and succulents may have got you wondering whether they are backed up by scientific proof.

Each succulent has different meanings in many cultures.
https://www.pexels.com/photo/assorted-color-flowers-2132227/

Scientific Significance

It was not until Columbus brought the first cactus back to the Old World (15th century) that scientific interest was generated in the plant. Since then, humans have discovered a lot about succulents in general.

Plants usually require quite a bit of moisture to survive. Succulents, on the other hand, can thrive without water for nearly three months. The secret lies in their thick, fleshy stems and leaves. Whenever there is moisture in the soil, they don't drink it all up in one go. They store the excess in their stems and leaves for a rainy day (irony intended), which is why they look so blown up like a balloon.

Visually, you can differentiate cacti from other succulents by the presence of areoles. They are bumps, like acne, on the stem that have spines shooting out of them. Another major part of the plant is the neck, which connects its stem to its roots.

Like other plants, cacti and succulents also undergo photosynthesis, but the process is somewhat reversed. In other plants, transpiration (carbon dioxide absorption) occurs during the day, but in succulents, the process is performed at night. It is done to avoid the loss of moisture stored in their stems.

Succulents are known to have various medicinal uses. Their antiviral and antibacterial properties prevent the occurrence of a number of infections and diseases. Do you tend to urinate more than usual? It could be due to prostate problems, which cacti can help to cure. They are also valuable for keeping your blood sugar levels in check.

Their fruit (if edible) contains vitamin C in abundance, ensuring increased immunity to most diseases. Many of their species are rich in antioxidants, and they also have anti-inflammatory properties.

Furthermore, the different fruits produced by cacti and succulents are generally edible, and the plants are also used as livestock fodder. But these are not the reasons they are popular inclusions in gardening and horticulture.

Entry into Gardening and Horticulture

For generations, humans have been fascinated by the appearance of cacti. From their proven medicinal uses to their probable healing powers, these plants have impacted several sects of people and cultures for centuries. To bring these scientific, spiritual, and visual benefits into their lives, they started looking for ways to bring succulents closer to home.

Since people staying somewhere other than the Americas couldn't gain easy access to the plants, it didn't take them long to figure out how to grow cacti and succulents in their private backyards or greenhouses and, to their delight, it was relatively easy to grow the plants and even easier to nurture them.

An Important Point to Note: While cacti are biologically just succulents, in the world of gardening and horticulture, succulents and cacti are two different entities. When you say that you're cultivating succulents, a fellow gardener will understand that you're growing anything but cacti, but if you're cultivating cacti, you should not say that you're cultivating succulents. Mention cacti specifically.

That said, back in the day, the spines of succulents and their general appearance may have kept many aspiring gardeners and horticulturists disinterested. They weren't grown as often as you would think before the 20th century, though in the last thirty years or so, their popularity has skyrocketed, especially after the advent of social media.

Today, showing off your homegrown succulents has become a trend, and several variants of cactus hashtags are being followed by hundreds of people every day. While there's nothing wrong with following a trend, you need not follow it blindly. Know what you're cultivating before you begin.

Taxonomy and Classification

Succulents are found in over 60 different plant families. The families that primarily contain succulents include Cactaceae, Crassulaceae, Agavoideae, and Aizoaceae. Cactaceae, as you might know by now, is your regular cactus plant with areoles and spines. It includes more than 125 genera and a little over 1600 succulent species. The only species that is naturally found beyond the Americas is the Rhipsalis baccifera. It is regarded to have once been native to the New World, too, but it may have migrated to Africa and Sri Lanka through pollination.

- **Crassulaceae**

Alternatively called the stonecrop family, Crassulaceae are likely to grow in any arid region of the world. They have succulent leaves that undergo photosynthesis via Crassulacean acid metabolism (CAM), a process similar to that of cacti with nightly transpiration. This family is known to consist of around 1300 species of succulents.

Crassulaceae (stonecrop family) are likely to grow in any arid region of the world.
Burkhard Mücke, CC BY-SA 4.0 <https://creativecommons.org/licenses/by-sa/4.0>, via Wikimedia Commons:
https://commons.wikimedia.org/wiki/File:Jard%C3%ADn_Bot%C3%A1nico_Mexico_City_82.jpg

- **Agavoideae**

This is a subfamily of the asparagus family of plants. Out of its 600 odd species, 300 are succulents. They are mainly found in the Americas but can grow anywhere in the tropics.

Agavoideae is a subfamily of the asparagus family of plants.
Pamla J. Eisenberg from Anaheim, USA, CC BY-SA 2.0 <https://creativecommons.org/licenses/by-sa/2.0>, via Wikimedia Commons:
https://commons.wikimedia.org/wiki/File:Agave,_Victoria_Regina,_Huntington.jpg

- **Aizoaceae**

This is one of the largest families of succulents, with nearly 2,000 different species. Many of these species are called ice plants since when you look at them under the sun, you can see them shine like ice crystals. Aizoaceae is also called the stone family because, in a land filled with rocks and pebbles, you will be hard-pushed to point out these ice plants.

Aizoaceae is one of the largest families of succulents, with nearly 2,000 different species. *Seweryn Olkowicz, CC BY-SA 2.5 <https://creativecommons.org/licenses/by-sa/2.5>, via Wikimedia Commons: https://commons.wikimedia.org/wiki/File:Aizoaceae_species_Greece.jpg*

With thousands of species to consider, it is sometimes hard to differentiate succulents from other plants. You need to keep a lookout for their individual characteristics.

Unique Characteristics

Succulents that are cacti are easy to identify. Does the specimen contain spines and areoles on a bloated stem devoid of leaves? That's definitely a cactus. It is recognizing *other succulents* that sometimes proves difficult!

- **Fleshy Stems**

Touch and squeeze its stem. Is it thick? Does it feel fleshy, like a plump fruit? It's the soft tissue of the succulents that help them efficiently store water.

- **Survival without Water**

Can the plant survive without water for several weeks? Water it once and wait for about seven days. Is it as fresh and healthy as it was on the first day?

- **Heat Resistance**

Most plants wilt or die in extreme heat. Does your plant seem to thrive in desert-like environments? If so, it is almost sure to be succulent because it probably has spines that are not visible to the naked eye, but they are big enough to keep the plant protected from excessive heat.

- **Thrive in Sandy Soil**

Unlike many other plants, succulents don't need a lot of water. Pot your plant in sand and water it once. Has it survived for a week or two?

Ecological Importance

Succulents don't just look good. They also positively impact the environment around them. Since they grow in a dry, hot climate where no other plants can survive, they are a great source of shelter for the fauna that inhabit the region. Many species of cacti can be given as fodder to domestic livestock, but they may be unpalatable and harmful for human consumption. The fruit of a few succulents is edible, but those of many others aren't recommended to be consumed raw.

The cochineal insect is one of the very few species that has adapted to survive on the water and nutrients of cacti. It feasts on the plant to its heart's content while protecting itself from predators by secreting carminic acid. This acid is used in the production of red dye (primarily used in food coloring and in the production of lipsticks). You won't generally find a single cochineal on a cactus since it reproduces and multiplies into several more while perched on the plant.

Succulents are decent absorbers of carbon dioxide, and since they usually live for more than 100 years, they are an excellent way to combat the growing global warming threat. Admittedly, a tree stores more CO_2 than a cactus, but the latter is easier to grow and care for and takes less space, so you will certainly reduce your carbon footprint with it.

Common Varieties of Cacti and Succulents and Their Benefits

Did you know there are more than 10,000 species of succulents in the world? Of these, there are nearly 1700 species of cacti, from the spiritually revered saguaro to the hallucinogenic peyote.

Succulent species usually differ in shape and form and in the presence or absence of leaves, spines, thorns, and areoles. Their stem may be a perfect sphere or as flat and thick as a pancake. But they all have two things in common, and that's the capability of storing water and their nativity in arid and semi-arid regions.

Not all species of succulents grow without the necessary climatic conditions. The types that are most preferred by gardeners and horticulturists for cultivation are the following:

- **Jade Plant**

The scientific name of the jade plant is Crassula ovata. It is originally from South Africa but is popularly cultivated throughout the globe. From afar, it looks just like a bonsai, but due to its thick leaves, it cannot readily be pruned like one. Its leaves are where it stores water.

The scientific name of the jade plant is Crassula ovata.
https://pixabay.com/photos/jade-succulent-green-plant-5220309/

Many succulent lovers prefer to cultivate this plant in abundance due to the ease with which it can be grown and also because it is low maintenance. It is best known for its healing properties and for curing diarrhea and nausea.

- **Aloe Vera**

Aloe Vera is famous for its skincare benefits, helping you look younger and wrinkle-free regardless of your age. It may reduce the recurrence of any persistent acne as well. The yellow-colored secretion from the plant, called aloe latex, may provide relief from constipation. But don't take it too often, or it may damage your kidneys and even prove fatal.

Aloe Vera is very easy to cultivate and nurture.
https://pixabay.com/photos/aloe-vera-succulent-cactus-botany-678040/

Aloe Vera grows naturally in the northeastern region of Oman and the eastern part of the United Arab Emirates. It can be cultivated anywhere except in places with prolonged snowy winters. It is primarily grown for ornamental purposes since the leaves are long, thick, and green with a jagged border. It is also simple to cultivate and nurture.

- **Burro's Tail**

Sedum morganianum is its scientific name. It is called burro's (donkey's) tail because its stem dangles from the pot like a donkey's tail. It belongs to the Crassulaceae family, and its natural habitat is in southern Mexico. Due to its unique hanging appearance, its pot is suspended from the ceiling to let its stems dangle freely.

It is called burro's (donkey's) tail because its stem dangles from the pot like a donkey's tail.
https://pixabay.com/photos/burros-tail-succulent-plant-6787172/

Burro's tail doesn't have any scientific or medicinal benefits per se, but any person walking by your house is bound to take a second look at the plant suspended gorgeously on your patio. It is also believed to cleanse your surroundings of negative energy, keeping the door open for positivity.

- **Bunny Ear Cactus**

Scientifically known as Opuntia microdasys, the bunny ear cactus's leaves are shaped like the ears of a bunny or the wings of an angel, depending on your visual preference. Apart from the obvious charming appearance of the plant, the gum secreted from its stem is used to create candles. Additionally, if you boil the stems in water, the soup can be blended with plaster to create a solid wall covering that doesn't chip off with time.

The bunny ear cactus's leaves are shaped like the ears of a bunny or the wings of an angel, depending on your visual preference.
https://unsplash.com/photos/SBKdiLOmylc

The bunny-ear cactus is a natural fog collector, meaning it can catch and store water from fog. It is indigenous to the desert regions of Mexico. A fair warning: the thin, tiny thorns on the cactus's stems can cause prolonged skin itching if touched. It is recommended to remove them soon after they grow.

- **Snake Plant**

As the name suggests, the snake plant is shaped like a slender, limbless snake. Scientifically called Dracaena trifasciata, it doesn't have any visible stems. Thus, its water storage and photosynthesis all occur in its leaves. It is one of the few succulents that can thrive in a relatively dark environment, making it perfect for potting indoors where there is little sunlight.

The snake plant is shaped like a slender, limbless snake.
https://unsplash.com/photos/iIuyXTcEBTI

For those who want more than just the snake plant's ornamental appeal, sleep in the room where it's placed. You will wake up to the smell of fresh, unadulterated air. Indeed, research has proven that the plant filters the air in its surroundings. Naturally occurring in West Africa, the natives often use it to soothe ear infections.

- **Prickly Pear Cactus**

The prickly pear cactus, or Opuntia, is among the most commonly cultivated cacti on the planet, thanks to its bright, semi-opaque flowers and delicious fruit. Its stems are often pad-like, similar to the bunny-ear cactus. The early growth of these pads is as beautiful as its flowers. It starts off as a small, pink-colored bud that grows into a large, flat, oval-shaped stem.

Opuntia, is among the most commonly cultivated cacti on the planet, thanks to its bright, semi-opaque flowers and delicious fruit.
https://unsplash.com/photos/DX_WW_J9Yh8

The prickly pear is native to many regions of the American continent, including Mexico, the United States, and the Caribbean islands. Its fruit is popular throughout its native lands and beyond. If you are eating a cactus fruit, chances are that it comes from the prickly pear.

Its probable medicinal uses include reducing inflammation, healing wounds, and even bringing down the risk of diabetes and obesity. The other well-proven uses of Opuntia are cochineal dye production, animal fodder, a leather alternative, fuel, and bioplastic production. Also, it is one of the easiest outdoor cacti to grow in your backyard.

As you may have noticed, two glaring benefits of cacti and succulents for gardening and horticulture are common to all plants.

1. They are easy to cultivate (among the easiest garden plants, in fact).
2. They are easy to maintain (water is rarely ever required due to their amazing storage capacity).

And you don't even need a lot of space to cultivate most succulents. A regular-sized pot or a small area in your yard is enough to make them grow and enhance your environment, not only visually but spiritually as well.

Chapter 2: Cacti and Succulent Selection: Which Should You Choose?

Now that you have learned about cacti and succulents, you may be wondering which plants are the better options for you. You should consider a few things first, like whether you live in a house with a garden or an apartment with a balcony. You should also consider your goal for the plant, your available resources, and the growing conditions.

This chapter provides you with all the information you need to decide between cacti and succulents.

Things to Consider Before Buying a Cactus or a Succulent

Some people just choose the best-looking cactus or succulent they find and then realize that the plant isn't really a good fit for their home. So, keep these things in mind before you make a purchase:

Garden vs. Balcony

Cacti and succulents can grow either indoors or outdoors. Although they prefer an indoor environment, many types can thrive outdoors. Before buying, check the tag for information. Make sure that the cactus gets enough sunlight every day. Don't place it in a shady area of your garden unless it thrives better in the shade. If you are going to keep it indoors, and it is a variety that likes sunshine, place it on a balcony or on a window ledge and make sure it is exposed to sunlight.

Cacti and succulents can grow either indoors or outdoors.
https://www.pexels.com/photo/clear-glass-table-decor-2100238/

Succulents prefer sunlight, so make sure they get enough every day. Orange, purple, and red succulents need a lot of sun exposure, so place them in the right spot at your home. Bright-colored succulents will be perfect for a house with a garden where they get to be in direct sunlight all day. If you live in an apartment, green succulents can be a good option for you.

Goals for Having a Cactus or Succulent

The second thing you should consider is your goal. Are you buying cacti or succulents for decoration or outdoor landscaping? If it's for decoration, make sure you have the right space for it in your home. However, if it's for creating a landscape, you will need a house with a garden. Since both cacti and succulents are decorative plants, this choice can be made using personal preference, but it is worth considering different plant heights.

If you use the plant for decoration, choose one with a pleasant scent. There are many cacti and succulents with flowery fragrances.

Plant Preference

Cacti and succulents can be prickly, flowery, colorful, or fruity. Before you make a choice, think about what you want. Perhaps a flowery cactus or succulent can be better for indoor decor, while the fruity ones are a better fit for your landscape. Colorful succulents can make a great addition to your garden, while colorful cacti are perfect for both. There is no right or wrong choice. This is about your taste.

Available Resources

Are you ready to add one of these plants to your home? Just like pets, plants are a huge responsibility, and they have needs that you should keep in mind. Make sure to prepare suitable pots for them. Find the right size, material, style, and drainage for your plant. For instance, succulents don't need a lot of water, so choose pots that have drainage holes. You should also prepare the proper position and make sure that your windows, balcony, or garden get enough light so your plants can thrive. This part will be explained in detail in the next chapter.

Growing Conditions

You must make sure you have the right growing conditions for your plant. Although both plants can grow in different climates, some types do better in cold weather than others. You should also consider the humidity level, temperature, and general weather conditions in your area. Choose the cactus or succulent type that is right for your climate.

Types of Cacti

This part will cover different types of cacti, together with giving helpful information so you can choose the right cactus for your environment and needs.

Bunny Ear Cactus

The bunny-ear cactus, also called polka-dot cactus, angel's wings, or bunny cactus, originates from Mexico. It is one of the most popular flowering types. Just like the name suggests, the plant's green parts are pretty similar in shape to a bunny's ears. It is a cute plant that is funny to look at, so it can be a lovely addition to your office and will surely put a smile on your face. It usually grows to about 23 inches, but it can take years to reach this height, as it is a slow-growing cactus.

Indoors vs. Outdoors
The Bunny Ear cactus grows indoors.

Home Decor or Landscape
Sometimes, pretty yellow flowers blossom on the cactus, which makes it a perfect plant to decorate your home.

Sunlight
It requires about eight hours of direct sunlight every day.

Water
Only water the bunny-ear cactus when the soil dries out.

Soil
Dry and well-draining.

Climate
The bunny-ear cactus can't tolerate cold weather and only does well in temperatures between 70 °F and 100 °F.

Old Man of the Mountain Cactus

You are probably wondering where this plant got its interesting name from. Well, it is covered with white and fine hair like an old man, making the name perfect for it. It is also called the Old Man of the Andes, and it originated in South America. At times, the hair can look like wool, giving it a distinctive look. The cactus spines are red and stiff and are easily distinguished against the white hair.

This plant is covered with white and fine hair, like an old man.
Jim Morefield from Nevada, USA, CC BY-SA 2.0 <https://creativecommons.org/licenses/by-sa/2.0>, via Wikimedia Commons: https://commons.wikimedia.org/wiki/File:Old_man_pricklypear,_Opuntia_polyacantha_var._erinacea_(38236654894).jpg

Indoors vs. Outdoors

The old man of the mountain cactus is an indoor plant. It is slow-growing and can reach two feet tall, but if it's planted in small pots, its growth will be limited.

Home Decor or Landscape

It is a decorative house plant.

Sunlight

It requires direct sunlight.

Water

Water it every 12 days during the summer and twice a month in the winter.

Soil

Well-draining soil.

Climate

Although it thrives better in heat, the plant can tolerate frost.

Walking Stick Cactus

The walking stick cactus has an interesting and particular appearance. It is a very thin plant, hence the name, and is covered with white, tiny, and sharp spines. The plant is also called the spiny cholla and the cane cholla and can reach a height of 4 feet. It grows a yellow fruit in the summer that you may mistake for a flower.

Indoors vs. Outdoors

The walking stick cactus can grow indoors and outdoors.

Home Decor or Landscape

With its unusual appearance, the walking stick is the ideal decorative plant. Since it can grow outdoors, it is also perfect for your landscape.

Sunlight

Direct sunlight.

Water

Water it every ten days while it's growing and once a month in the winter.

Soil

Well-draining soil.

Climate

The plant thrives in dry weather.

San Pedro Cactus

The San Pedro cactus is a green plant with white and tiny spines. It can grow to a height of 20 feet, and white-scented flowers bloom on its pads at night.

The San Pedro cactus is a green plant with white and tiny spines.
https://www.pexels.com/photo/green-san-pedro-cactus-10109786/

Indoors vs. Outdoors

The San Pedro cactus can grow indoors or outdoors. If you plant it in a small pot, it will limit its growth.

Home Decor or Landscape

With its beautiful, scented flower and eye-catching green color, the San Pedro cactus is a great decorative plant.

Sunlight

This cactus requires direct sunlight to grow and doesn't do well in the shade.

Water

Only water it when the soil is dry.

Soil

Water-permeable soil.

Climate

It can grow in warm temperatures but won't survive the frost.

Brazilian Prickly Pear Cactus

The Brazilian prickly pear cactus is another interesting-looking plant. It has rounded and flat pads with flowers blooming over them. It also has small spines growing all over it and orange and yellow flowers blooming. The flowers transform into orange, purple, and yellow bulbs that resemble a pear, hence the name. The pears are edible and have a wonderful fruity fragrance. The Brazilian prickly pear cactus can reach 30 feet in height.

Indoors vs. Outdoors

The Brazilian prickly pear cactus can grow indoors or outdoors.

Home Decor or Landscape

It can be a decorative houseplant.

Sunlight

This cactus needs direct sunlight, but if you keep it indoors, make sure to place it on the balcony or near a window to get indirect bright light.

Water

Water the cactus when the soil dries out.

Soil

Well-draining soil.

Climate

The plant thrives in the summer and warm climates but can wither in extremely hot weather. It can tolerate the cold weather, provided you keep it dry.

Strawberry Cactus

The strawberry cactus is also called the hedgehog cactus because of its tiny and spiky stems. It is a desert plant with beautiful deep pink spring flowers and can grow up to 3.5 inches.

Indoors vs. Outdoors

The strawberry cactus is an outdoor plant.

Home Decor or Landscape

The strawberry cactus will be perfect for your landscape if you live in a dry environment. The cactus's spines can be gray, white, brown, and yellow, which can make your garden look colorful.

Sunlight

This cactus requires either full or indirect sunlight.

Water

Water it once a month during the winter and twice a month in the summer.

Soil

Well-draining soil.

Climate

Hot, dry climate.

The Old Man Cactus

The old man cactus is covered with thicker hair than the old man of the mountain cactus, and it looks like it is wearing a fur coat. Beneath the thick white hair are tiny, sharp spines. It can reach a height of fifty-two feet and can grow white, yellow, or red flowers, but it can take 10 years for the flowers to bloom.

Indoors vs. Outdoors

The old man cactus can grow indoors and outdoors.

Home Decoration or Landscape

The cactus can be a houseplant, as its wooly appearance makes an interesting decoration. It can also be a great addition to your garden.

Sunlight

It requires direct sunlight but thrives better in light afternoon shade.

Water

Only water it when the soil is dry.

Soil

Well-draining soil.

Climate

The cactus thrives in warm weather and can't tolerate the cold.

The Queen of the Night Cactus

The queen of the night cactus blooms into stunning white flowers once a year, at night, which is how it gets its interesting name. It is also called Dutchman's pipe cactus. It is native to South America and Southern Mexico. It can grow up to twenty feet tall. Its flowers grow over the cactus's green stems and have a subtle, soft scent. They grow 11 inches long and five inches wide. The flowers' large size gives them a unique feature.

The queen of the night cactus blooms into stunning white flowers once a year, at night, which is how it gets its interesting name.

https://pixabay.com/photos/queen-of-the-night-cactus-flower-7279135/

Indoors vs. Outdoor

The queen of the night cactus thrives in an indoor environment.

Home Decoration or Landscape

Thanks to its beautiful, large flowers, it can be a decorative houseplant.

Sunlight

The plant needs direct sunlight to grow, but only for a few hours in the morning; it then requires indirect light.

Water

Only water the plant when the soil dries.

Soil

Airy and well-drained soil.

Climate

Semi-desert and tropical climates. It can tolerate extremely cold weather for a short amount of time. However, if it is exposed for a prolonged time, it will wither and die.

Old Lady Cactus

The old lady cactus is round and covered with tiny spines. It can reach a height of 20 inches and blossoms into reddish-purple flowers that grow in a circular shape on top of the plant.

Indoors vs. Outdoors

The old lady cactus can grow indoors and outdoors.

Home Decoration or Landscape

It is a decorative plant.

Sunlight

It requires direct sunlight but can also thrive in indirect sunlight.

Water

Water once a month during the winter and every week in the summer.

Soil

Mixed soil.

Climate

The cactus thrives in warm weather but can struggle in extreme heat. Keep it dry in the winter.

Christmas Cactus

Christmas cactus originated in Brazil, and it is one of the most popular indoor plants in the world. It has long, flat stems and no leaves. In the winter, it blossoms into red, yellow, pink, and white flowers.

The Christmas cactus originated in Brazil.
https://pixabay.com/photos/plant-leaf-nature-cactus-3101751/

Indoors vs. Outdoors
The Christmas cactus is an indoor plant.

Home Decoration or Landscape
Thanks to its trailing stems, you can place it on surfaces or hang it from baskets.

Sunlight
It requires indirect sunlight to grow.

Water
Water the cactus every week or two weeks.

Soil
Acidic soil.

Climate
The plant thrives in a humid environment.

Types of Succulents
Now, you will learn about different types of succulents.

Starfish Succulent
The starfish succulent owes its name to its beautiful flower that resembles a starfish. The flower is reddish-purple, deep red, or orange. However, its scent doesn't match its beautiful colors as it smells like rotting meat to attract insects. It can grow to 12 feet tall.

The starfish succulent owes its name to its beautiful flower that resembles a starfish.
Red Bead Dragon, CC BY-SA 4.0 <https://creativecommons.org/licenses/by-sa/4.0>, via Wikimedia Commons: https://commons.wikimedia.org/wiki/File:Stapelia_Grandiflora_Flower.jpg

Indoors vs. Outdoors

The starfish succulent can grow indoors and outdoors. Since it has an unpleasant odor, it's better if you keep it outdoors.

Home Decoration or Landscape

It can be used as a decorative plant.

Sunlight

It requires full and direct sunlight and can also thrive in bright indirect light.

Water

Water it once a month during summer and spring and every two months in the fall and winter.

Soil

Well-draining soil.

Climate

The starfish succulent thrives in warm weather and can tolerate freezing weather but only for short periods, and the soil must be kept dry.

Afrikaans Succulent

Afrikaans succulents are pretty plants with big and flat leaves. They grow upward and have yellow or white flowers. Interestingly, when the leaves are fully grown, they split out to look like a peace sign.

Indoors vs. Outdoors
Afrikaans succulents can grow indoors.

Home Decoration or Landscape
Afrikaans succulent is a perfect decorative plant.

Sunlight
The plant requires direct sunlight.

Water
Water the succulent when the soil dries out.

Soil
Loose, grainy soil mixture.

Climate
Warm climate.

Aloe Vera Succulent

Aloe Vera is one of the most popular succulents since it is known for its cooling effects. The aloe Vera gel is extracted from the leaves and is used in treating many skin conditions like rashes and sunburns. It is also a pretty plant with a stunning bright green color.

Indoors vs. Outdoors
Aloe Vera can grow indoors and outdoors.

Home Decoration or Landscape
It is a decorative plant.

Sunlight
The plant needs direct sunlight to grow.

Water
Water when the soil dries out.

Soil
Mixed soil with lava rock and perlite.

Climate

It thrives in hot and dry climates and should be kept indoors during the winter.

Jade Succulent

The Jade is another popular succulent since many people use it to decorate their homes. It is a slow-growing plant with a lifespan of 70 years. It can grow up to six feet tall and has woody and thick stems with oval-shaped leaves.

Indoors vs. Outdoors

Jade plants can grow indoors and outdoors.

Home Decoration or Landscape

It can be a houseplant and adds a lovely ambiance to your home.

Sunlight

It requires full and direct sunlight.

Water

Water when the soil dries out.

Soil

Well-drained and acidic soil.

Climate

It thrives in a warm environment and can even survive in the frost.

Mexican Snowball Succulent

The Mexican snowball is also called Mexican gem and white Mexican rose. The plant is known for its silver-green and blue-green leaves that have a rosette shape. It originated in Mexico and usually thrives in a semi-desert environment.

Indoors vs. Outdoors

Mexican snowballs can grow indoors and outdoors.

Home Decor or Landscape

People from all over the world decorate their homes with this succulent.

Sunlight

It requires direct sunlight.

Water
It doesn't need much water, so only water it when the soil dries out.
Soil
It requires well-drained and acidic soil.
Climate
The plant thrives in hot and dry climates but will struggle in humid weather, and it can't tolerate extreme winters.

Crown of Thorns Succulent

The crown of thorns can grow up to six feet tall. However, if you plant it indoors in a small pot, it will limit its growth, and it will only reach two feet in height. It has green and thick leaves, but they are covered in thorns, which is where the name comes from. The plant blooms into producing white, yellow, pink, orange, or red flowers.

The plant blooms into producing white, yellow, pink, orange, or red flowers.
ಕಿಖಿಞ್ಚಿಗೂ, CC BY-SA 3.0 <https://creativecommons.org/licenses/by-sa/3.0>, via Wikimedia Commons: https://commons.wikimedia.org/wiki/File:Euphorbia_milii_-_%E0%B4%AF%E0%B5%82%E0%B4%AB%E0%B5%8B%E0%B5%BC%E0%B4%AC%E0%B4%BF%E0%B4%AF_07.JPG

Indoors vs. Outdoors
It can grow either indoors or outdoors.
Decorative or Landscape
It is one of the most popular decorative houseplants in the world.

Sunlight
It can thrive in full or partial sunlight.
Water
Water when the soil dries out.
Soil
Well-drained and neutral soil.
Climate
Warm and dry weather.

Chocolate Soldier Succulent

Chocolate soldier succulent, also known as the panda plant, is a pretty and attractive plant that you find in many households. It has pale green leaves with brown spots and is covered in white-gray fuzz. It is slow-growing and can reach 2 feet in height.

Indoors vs. Outdoors
Chocolate soldier can thrive outdoors in warm weather but prefer an indoor environment.
Decorative or Landscape
It is an attractive plant that is used to decorate many homes.
Sunlight
The plant will wither in direct sun exposure; it only thrives in indirect sunlight.
Water
Water when the soil dries out.
Soil
Well-draining, neutral, and acidic soil.
Climate
It can't tolerate extremely cold weather and thrives in warm to hot weather.

Flaming Katy Succulent

The plant has green, oval-shaped leaves. Under direct sunlight, some varieties have red leaves. It has small white, salmon, orange, yellow, pink, and red flowers.

Indoors vs. Outdoors

It thrives outdoors during the summer but prefers an indoor environment in the winter.

Decorative or Landscape

The flaming Katy can be an ornamental houseplant.

Sunlight

When planted outdoors, it requires full sun exposure or partial shade, while indoors, it needs indirect light.

Water

Water it every few weeks.

Soil

Sandy, well-drained, acidic, and neutral soil.

Climate

It thrives in warm weather and can't tolerate the cold.

Ponytail Palm Succulent

The ponytail palm succulent is an extremely popular office plant. It can grow up to thirty feet tall outdoors. It can grow to be a large tree, bigger and taller than many homes. When planted indoors, it only reaches six feet tall. It originated in Central America, and it can take years to grow. It has a long lifespan, so it will be with you for a long time.

Indoors vs. Outdoors

It can grow indoors and outdoors. Plant it indoors if you want to restrict its growth.

Sunlight

It requires direct sunlight or indirect sunlight exposure.

Water

Water it every two weeks during the summer and every month in the winter.

Climate

It thrives in warm weather and can tolerate extreme cold for a short period.

Quiz

Now that you have become familiar with different types of cacti and succulents, take this quiz to determine which one is the right fit for you.

1. **Do you live in a warm or cold climate?**
   - Warm climate
   - Cold climate
2. **Do you have a garden in your home for outdoor plants?**
   - Yes
   - No
3. **Do you have a window or balcony for indoor plants that require sunlight?**
   - Yes
   - No
4. **Do you prefer flowering plants or ones that are covered with white hair?**
   - Flowering plants
   - White-haired
5. **Do you think you can provide the appropriate growing conditions for your plant?**
   - Yes
   - No

Now check your answers, then go over the list again and narrow down your options. You are now ready to make the right choice for your needs.

There are many types of cacti and succulents. Some grow indoors, some grow outdoors, while others can thrive in both. Some plants can be a nice decorative addition to your home, while others can be a perfect plant for your garden. There isn't a right or wrong choice. You just need to find a plant that fits your imagination and living space.

Make sure you consider all aspects before making a choice, and don't ignore the plant's scent. Some have an unpleasant odor, while others can be flowery or fruity. Although it is tempting to choose one based on its attractive shape or colors, it may not thrive in your environment. So don't make a rushed decision.

Chapter 3: Planting and Arranging Your Cacti and Succulents

So, you visited your local plant distributor, picked out a beautiful and unusual assortment of cacti and succulents, and took them to their new home. Now what? There are so many variables of gardening techniques to consider for these exquisite plants. Should you plant the succulents and cacti together? Should you plant them separately? Do you use one pot or more? What soil preparations should you consider? How will you arrange them - individually and collectively? This and so many other questions will be addressed and answered throughout this chapter.

To avoid any confusion as you move along, remember all cacti can be considered succulents, but not all succulents are considered to be cacti. They are similar in many ways: they are both resistant to heat and lack of water, and both have thick structures. However, if you explore their differences, cacti are known for the spikes that emerge from the plant's surface that reduce the air flowing around it to optimize the water supply and protect the plant.

Another major difference between them is the photosynthesis technique. With cacti, the process takes place through the stem rather than the leaves, as happens in succulents and other plants. This is why cacti require much less water to survive and are able to withstand harsher environments.

These resilient and undemanding plants can add color and alter your garden landscape into an unforgettable aesthetic. Taking on the challenge of mixing and matching plants can be unnerving. To give you a hand with this daunting task, you'll find a step-by-step guide on what to do and what to avoid as you get further into the chapter.

Indoors vs. Outdoors

Transferring your cacti and succulents indoors is a possibility as long as you are able to replicate the environment they come from. This usually involves an abundance of light, well-drained and dry soil, and cool temperatures inside.

- Placing your plants near a window should ensure light exposure and lower nightly temperatures.
- Some cacti tend to flower and thrive in an indoor environment, though don't be fooled by unscrupulous sellers trying to sell you plants not suited to an indoor environment.
- Among the cacti that flower indoors are Lobivia and Rebutia.
- Lower humidity indoors tends to agree with some of the species; levels that are no lower than 10% and no higher than 30%.

Outdoor succulents tend to grow and prosper in the summer sun.

- As the spring rolls around and the weather becomes hotter, start by moving them into a shady area of your landscape or garden.
- Slowly but surely, start the transition into sunnier locations as the plants adapt to the sun.
- Try to place them in areas where they won't be exposed to the sun's direct heat from around mid-morning to midday and slightly after (from 11 a.m. to 3 p.m.).
- Once you've made the transfer outdoors, cacti may very well not require watering – *but do check them regularly.* As for succulents, they may need a little bit more moisture than the cacti.

Planting Locations

First, start by deciding whether you're planting them individually or together and what is required when choosing the location either way.

Planting Cacti Alone

The best time to plant cacti is in spring and summer. Cacti can be planted inside or outdoors as long as there is enough light for them to thrive. The next issue is the kind of soil. The main thing to consider with the soil you choose is that it should be well-draining.

If you're going to choose to plant them indoors, make sure that they are either facing south or west windows. If you're looking to move your indoor plants to your outside landscape, start by placing them in a protected area at first. Gradually, as they get used to the outside environment, start moving them to areas with more sun.

When considering containers, go for small and shallow ones. Cacti have shallow, slow-growing roots, so they don't need deep containers.

Consider adding small pebbles and rocks to the bottom of your container or an elevated bed. Make sure that the potting mix you use is specifically designed for cacti, as most other mixes may hold more water than needed, which could lead to problems.

The best time to plant cacti is in spring and summer.
https://www.pexels.com/photo/three-potted-cactus-plants-1903965/

Planting Succulents Alone

Succulents are slightly more sensitive than cacti, and that does not diminish their resilience. Make sure you are familiar with the kind of succulents you're planning to grow and whether they are more suitable for indoor or outdoor environments. Some may not withstand extreme conditions like others.

Like cacti, soil drainage is a vital factor when planting succulents. To improve that quality, try adding pumice or sand to your soil. Succulents do not fare well in wet soil, and materials that can help loosen the soil are small gravel, surface, and perlite.

Like cacti, succulents are better suited to smaller and shallower pots. When planting, make sure not to backfill the soil. Allowing for the gaps of air ensures the growth of new roots near the surface, where they are able to breathe.

Make sure you are familiar with the kind of succulents you're planning to grow and whether they are more suitable for indoor or outdoor environments.
https://www.pexels.com/photo/top-view-photography-of-three-succulent-plants-2516658/

Planting Cacti and Succulents Together

If you choose to plant these two plants together, remember that they each have different watering requirements. Cacti can survive in much drier environments than succulents, while succulents cannot thrive in drought. So, to make sure they both have their needs met, choose a patch of land with direct access to the sun and soil that drains properly.

Make sure to separate the two beds housing the plants. While cacti prefer sandy soil, succulents favor a slightly heavier environment with more oxygenation and a slightly higher demand for irrigation.

As mentioned before, not all cacti are the same, and neither are the succulents. Make sure that the types you're planting together can tolerate the weather and environmental conditions in the area.

When choosing the container, go for a bigger one to make room for both separate beds, allowing room for the growing roots. Make sure your container has holes to drain any excess water and isn't made from material that absorbs it, so go for something like ceramic containers.

Make sure to separate the two beds housing the cacti and the succulents.
https://pixabay.com/photos/cacti-plants-succulents-pots-1845159/

Preparing the Planting Area

When preparing your area for planting, remember that cacti and succulents both do well in mildly acidic soil, with a sandy composition for optimal drainage of excess water.

You can perform a soil test before planting, keeping in mind that the optimal acidic PH balance should range from 5.5 to 5.6.

Depending on the results you get, if they are not within the correct range, you can easily amend the situation by adding sand or humus/girt.

For cacti, having a blend of about 50% sand, 25% humus, and 25% grit would be ideal. Remember, though, that succulents require a heavier growing medium, so maybe increase the percentage of the soil in their section of the bed.

Consider adding organic material to the mix, such as manure or compost.

A viable test for the soil mixture is to wet it and squeeze it tightly with your hands. When you open your hand, the soil should fall apart.

Drainage holes are essential to avoid rotting and withering plants.

If you're planting outside or inside, make sure that the soil is loose and properly aired using a gardening fork. After you've done that, spread a layer of mulch on the surface to reduce weed growth and keep it moist.

It is advised to start preparing the soil 14 days prior to planting. For succulents, try choosing areas with a proper sun-to-shade ratio.

Clean out any weeds, debris, or overlooked pests from previous plantations from your planting area. These unwanted additions to the soil can hinder plant growth or cause irreparable damage.

When placing rocks around your cacti and succulents, you will need to be strategic. Adding a top layer of smaller decorative rocks doesn't just look good but also allows for the soil to retain the heat. This sort of heat retention comes in handy during colder weather to support the sun-loving plant.

Big rocks also act as deterrents to predators, such as medium-sized animals (dogs and deer), from stepping on them and crushing them.

Make sure the ground around your outdoor garden is level to avoid water pooling.

If the area where you are located is known for cold winters with excessive moisture, consider moving your cacti and succulents under an overhead roof or under transparent/translucent coverings to keep them dry.

Planting Techniques

Whether it's growing from seeds, propagation, plant cuttings, or transplanting succulents, there are specific steps you need to follow to be successful.

Transplanting Cacti

- Make sure you have thorn-proof gloves or even kitchen tongs to protect your hands from cacti thorns.
- Another method of keeping safe is wrapping newspaper layers around the prickly bits of the cactus that you plan to move. Five or six layers should keep you safe from any injuries.
- When attempting this, make sure the weather is appropriate. The sunnier, the better, so springtime to early fall is a good time.
- Check and make sure the roots of the plant are completely dry before you move it.
- Dig a hole in the ground around the same depth and width to match the one the potted cactus is in.

- To loosen the soil in the original pot, tap lightly on and around it.
- Carefully and gently remove the cactus from the pot with a shovel with as little disturbance to the roots as possible.
- If the sand mixed in with the soil falls apart in the move, use it to backfill the hole to which you're adding the cactus.
- Use tongs when handling the cacti to avoid injuring yourself, as gloves offer little protection against spines.
- Backfill the hole with the appropriate mix of soil for cacti (mentioned above) until it is right above the roots.
- Add a layer of gravel or lava granules on top of your soil around the cactus.
- Leave the plant for a week before attempting to water it.

Propagation of Indoor Cacti

The use of pups (baby cacti or offshoots of the cacti) not only allows you to increase your succulent population – but also improves the overall health of the parent plant.

Offshoots usually occur at the base of the cacti, and they share nutrients and water with the main plant. Sometimes, they may occur on the body of the cactus, whether that be the stem or the pads.

Here is how propagating works.

- Make sure you have protection gloves, cactus soil mix, alcohol pads or spray, a sharp knife, rooting hormones, and a pot, of course.
- Make sure you clean the knife with the alcohol and then leave it out to dry.
- Put on your protective gloves to avoid injuries.
- Choose the offshoot or pup you wish to propagate, and using the knife, cut at the base where it connects with the parent cactus. Angle the knife at 45 degrees so that the resulting cut firms up quickly before the rot sets in.
- Do not pot the pup right away! Leave it out to dry completely for up to a week, allowing it enough time to callus or dry out.
- Add the cactus potting mix to the pot.
- Insert the cut end of the offshoot into the rooting hormone and then press it gently into the top of the pot.

- Leave it out in the indirect sunlight and add droplets of water to it often.
- Within four to six weeks, your cactus should have grown strong roots.

Growing Cactus from Seeds

Growing from seeds can take a long time and requires a lot of patience. The seeds used to plant cacti come from the flowers that sprout from an adult plant.

Because some cacti never really develop any flowers, opting to buy your seeds from a commercial shop may be the only option.

Cacti seeds may require stratification. This is a process in which a simulation is performed to make it seem as if the seeds have experienced winter before starting the planting process.

Stratification occurs by placing the seeds in a wet brown deposit that doubles as soil called "Peat." Peat is formed from the partial decomposition of vegetable matter in wet, acidic conditions, muddy ground called "bogs," and flooded areas of land called "fens." Peat is often dried out and repurposed for gardening and fuel needs.

After placing your seeds in moist peat, they are placed in your cooler for about four to six weeks until they crack open.

After this process is concluded, make sure you have a cactus soil mix available. Plant the seed in the pot with the mix at an equal depth and width. Cover up the pot with plastic after watering it lightly and place it in a bright location but out of direct harmful sun rays.

Sprouting usually occurs within three weeks, and the covering can be removed during the day after that. After about six months, little seedlings are ready to be repotted.

Planting Cacti Outside

If you're going to choose to have your prickly plants outside, here are the steps to follow:
- Make sure the soil is well prepared with the right cactus mix.
- Make sure the hole you're digging in the ground is as equally deep as it is wide. It should be about one and a half times as wide as the cactus's root ball or stem.

- Set the plant facing north. The north side of the plant usually has an obvious mark on it. If there isn't a mark, make sure you ask about it in the plant store prior to leaving. Why should you do that? Because the plant's south side is most likely placed more in the sun and develops tougher skin than the north side. Incorrect placement may cause the north side to receive more sunlight than it can withstand.
- Backfill the hole with the soil mix and pat it gently afterward.
- Add small amounts of water.
- If you are replanting a cactus bred in a greenhouse, cover it up with a light cloth to shade it from direct exposure to the sun for a couple of weeks.

Planting Cacti and Succulents Together

Before planting, make sure that you've chosen the right combination of cacti and succulents that complement each other visually and can be sustained together in a common environment.

If you're going for a cactus that can thrive in bright sunlight, don't choose a succulent that can easily burn from sun exposure, such as the string of hearts or the snake plants.

Make sure your container is wide enough for both plants to exist comfortably together and has enough room for root growth. Leave about six inches of space between them to allow a healthier co-existence.

And now, on to the repotting process:
- Make sure you've watered your cacti and succulents two days prior to transplanting to the new pot. As the plants are placed in new soil, they need time to adjust to their new environment, so it's best not to water them at all for the first two weeks after repotting. Watering beforehand allows your plants to take their fill of water, and the wet soil makes it easier to remove the roots from the pot.
- Prior to adding the cactus or the succulent, take a closer look at each plant. Make sure that you've checked for any signs of rot, pests, or disease indications. If the plants are left without treatment and joined together in one pot, the infected plant will spread its illness to the rest. Cut off any damaged parts and prune the plants if needed.

- When arranging the plants, pick the design you like. You can go for taller plants at the back and shorter ones at the front, for example.
- You can arrange the planter from the middle. Maybe add your biggest plant and rotate the rest around it.
- A vital thing to remember is that when planting your cacti, leave either one-third or half of the root exposed and unburied and cover it up a bit later on. For succulents, you need to make sure to cover all the roots, as they need more moisture than cacti. When you follow this technique and leave some of the cacti roots above the soil, you avoid overwatering them since they need less water than succulents.

Arranging the Plants for Visual Harmony

There are several ideas to explore when arranging your resilient plants individually or together. Each person has their own taste, and each arrangement can vary dramatically according to the weather and what you're pairing it with.

Here are a few simple ideas for your indoor and outdoor arrangements:

- If you're planting only succulents, try adding different colors and shapes close together to create a tapestry.
- Suppose you're looking for specific themes for your garden. In that case, you can create your botanical masterpiece with the inspiration of life under the sea. Pick out colors that match undersea plants and creatures and start arranging them in the form of coral reefs.
- Try writing words with your plants by planting them in wooden frames of letters.
- You can try adding your plants to baskets instead of pots. For example, forest cacti are famous for their huge flowers, which makes them ideal for individual displays.
- You can add the succulents in two varying sizes and colors in a landscape with a pathway surrounded by white pebbles.
- You can add crevices of rocks to your garden and make it look like the cacti and succulents are sprouting directly from them.
- You can emphasize the look of the garden by raising the soil where your plants are situated to draw the eye toward them.

- Consider filling the nooks in your stone walls with soil and tucking your plants into them. Maybe throw in colorful stones to make the colors more vibrant.
- Try creating a Zen garden. Instead of small rocks and pebbles, add big, flat ones as a base for your cacti pots. Spread smaller rocks between the pots to make it resemble a Japanese garden. Try mixing and matching the color of the plants with the pots and rocks to create the aesthetic you're looking for.
- You can improvise the design of the fence around your home. Instead of wooden planks, you can replace them with cacti. You can plant your succulents along the pathways to create a visual separation between the different areas of your garden. Make sure you add a warning sign next to your plants to avoid unintended injuries.

Chapter 4: Watering Wisely: The CORRECT Way to Water Cacti and Succulents

Cacti and succulents are wonderful plants. Their beauty and numerous benefits make them a pleasing plant choice in the home, but often, there is a preconceived notion about how you should water and maintain these plants. Many think succulents and cacti require little to no water, more water than houseplants, or just as much water as houseplants. Some other common assumptions regarding watering cacti and succulents include using regular potting and well-drained soil and watering during dormant seasons. Some of these assumptions are indeed true. However, it is better to be certain to get the best out of your cactus and succulent plants.

Properly watered plants brighten your home in any season, clean the air around you, strengthen your immune system, and improve your mood. Not getting it right can cause enough damage to undo all of your previous efforts, and once it's done, the difference in your plants is obvious. Therefore, you must know how to water your plants correctly.

This chapter will provide information on how to water your succulents and cacti plants properly to avoid over or under-watering. Additionally, you will learn different techniques and the right tools to use. When you put the information in this chapter into practice, you will become a pro on the *dos and don'ts* of watering your cacti and succulents in no time.

Unique Water Requirements of Cacti and Succulents

Cacti and succulents have distinct water needs, which sets them apart from most common plants. Their watering process has helped them flourish in arid areas, semi-arid regions, and areas with little water availability. Therefore, understanding their distinct and peculiar water requirements when you want to grow them in your home is essential to successfully cultivating and caring for these fascinating and resilient plants. Here are the requirements for watering cacti and succulents.

Cacti and succulents have distinct water needs.
https://pixabay.com/photos/cactus-watering-can-houseplant-4161380/

1. Infrequent Watering

Infrequently watering your cacti and succulents is vital to caring for them. Since these plants don't do well in wet conditions, ensure that watering is only done when the soil is completely dry. Avoid initiating a regular watering schedule for your cacti and succulent plants.

Infrequently watering your plants is not the same as neglecting them. When you neglect your plants, they will weaken and whither. Therefore, water your succulents and cacti every fourteen days in summer and every thirty to fifty days during winter.

2. Watering Techniques

Like the watering frequency, the watering technique of succulents and cacti differs from that of other plants. Most plants are watered from top to bottom, but succulents and cacti are watered from the bottom. This makes sure that only the roots get wet and the upper plant stays dry.

3. Drought Tolerance

It's no news that succulents and cacti thrive in areas that some regular plants won't be found in because of their adaptability to arid areas known for their dry and harsh nature. Succulents and cacti possess special tissues for storing water, such as the fleshy stems and leaves, making them tolerant to drought. Thus, you won't be helping them by keeping them wet or damp through a morning and evening watering schedule.

4. Well-Draining Soil

A soil that drains effectively or allows water to flow through freely is another key requirement for watering succulent or cactus plants. A well-draining potting mix should contain materials such as sand, pumice, or perlite, as these elements will prevent water from pooling around the base of your plant. Also, they are a great choice because they help aeration of the soil, which is essential for root health.

5. Rainwater and Water Quality

Rainwater is another requirement for watering succulent or cactus plants. Still, because it isn't always accessible, you can use regular tap water after letting it sit for at least a day. By letting the tap water sit, chlorine and other chemicals evaporate, allowing you to use it to water your plants.

You can also make use of greywater recycling in place of rainwater. Greywater recycling is when household water from the sink, shower, or other parts of the house is reused after appropriate filtration and irrigation.

Furthermore, you can use distilled water. However, distilled water should only be used occasionally because it doesn't contain essential minerals.

6. Seasonal Adjustment

Cacti and succulents are more active during growing seasons like summer and spring. In these seasons, they require extra water. However, more water does not translate to the same amount of water you would use for other plants in your garden. It just means watering should be done more often than normal. Conversely, during seasons like fall or winter, which is their dormant phase, very little water is needed.

7. Observation and Adaptation

There are different species of cacti and succulents, and the distinction in this plant type also means a distinction in their requirements. The specific water requirements of these plants can influence how well they perform. So, after getting them, watch them closely in the first few weeks to work out how much water each plant needs.

8. Plant Environment

The environment where you plant your succulent or cactus also influences how much water they need. For instance, they tend to get dryer when planted in containers rather than directly in the ground. Therefore, if your plant is in a container, it will require more frequent watering than when planted on the ground.

In summary, the special water requirements of cacti and succulents revolve around their ability to store water, their tolerance for dry conditions, and the need for infrequent watering and well-draining soil. Understanding these specific needs and providing appropriate care can help these remarkable plants thrive in your home or garden.

Risks of Overwatering and Underwatering

In the process of watering, many owners of succulents and cacti are unable to get the perfect balance between over and under-watering. This section discusses the risks of under-watering and over-watering these plants.

Risks of Overwatering Succulents and Cacti

Cacti and succulents are built to store water in their tissues, so it's crucial not to overwater them. Overwatering is one of the most common causes of issues with these plants. It's safer to water less frequently when in doubt. Proper watering practices will help your cacti and succulents thrive and prevent potential problems caused by excessive moisture. The following are the dangers of overwatering:

- **Root Rot**

One major effect of overwatering is rotting the roots. Since water is applied to the base of the plant and pools around the base, the roots can be affected by overwatering. When the roots sit in water for too long, they become soft and deteriorate. This rotting process interferes with their natural function, preventing them from absorbing nutrients from the soil properly.

- **Bacteria Problems**

Another problem of overwatering is bacteria multiplication. Microbes multiply in large numbers when plants deteriorate and rot in wet environments.

- **Fungi Problems**

Fungi infestations are a direct problem of overwatering. This problem frequently accompanies root rot, which creates the perfect environment for the emergence of these microorganisms. Fungi infections, such as gray mold and powdery mildew, cause disfiguration and sometimes even death of these plants if the problem is not treated on time.

- **Insect Infestations**

Insect infestation is another key risk brought on by overwatering your cacti or succulents. Certain insects, like thrips or mealybugs, prefer wet areas and love to live on the wet parts of plants. When succulents and cacti are overwatered, bugs or pests infest the cacti or succulents in their weakened state and start reproducing. These insects lay their eggs deep within the plant's tissue, and as they mature, the larvae feed on the delicate interiors of the plant's leaves, flower buds, or stems. After that, infestation takes over, making it nearly impossible to eradicate.

Risks of Underwatering Succulents and Cacti

In the process of watering, many owners of succulents and cacti are unable to get the perfect balance between over and under-watering.
Sreyasvalsan, CC BY-SA 4.0 <https://creativecommons.org/licenses/by-sa/4.0>, via Wikimedia Commons: https://commons.wikimedia.org/wiki/File:Drought_in_a_Lake.jpg

- **Cactus Shriveling**

In the process of preventing over-watering, you could end up under-watering, leaving you with a shriveled cactus plant. Shriveling happens when there is very little water and hydration in your plant's tissues. This dehydration also leads to shrinkage and distorted growth in your plants.

- **The Yellowish Appearance of Cactus**

When cacti are under-watered, they begin to turn yellow. Plants turn yellow when they find it difficult to produce chlorophyll due to damage to the roots of the plants.

- **Cactus Wilting**

One of the many roles of water in cacti is helping maintain the form of the plant. Without enough water within these plants, they start to droop and wilt. However, wilting is more prevalent in cacti or succulents with a more upright and thin posture, which is noticeable when they become soft and unable to hold themselves up.

- **Leaves with Dry Brown Spots**

Dehydration of these plants can lead to brown spots or patches on their leaves. When dehydrated and too close to a heat source or facing direct sunlight, they become susceptible to heat damage and sunburn, which appears as dry brown patches.

- **Brown Patches on the Edges and Tips of Plants**

The tissues of the cactus are meaty, and when the plant is under-watered, brown patches appear around the edges and tips of the plant.

- **Curling of Leaves**

When your cacti or succulents are under-watered, the plant's coping mechanism is to curl up. They do this to save whatever water they have left in them, causing their roots to eventually rot.

- **Weak Spines**

Lack of proper hydration in these plants can result in brittle, weak, and fragile spines. When their spines become weak, they fall off on their own or when touched. As in the case of curled leaves, weak spines can also be a symptom of overwatering.

- **Brittle Roots**

When plant roots are not getting as much water as they should, they become brittle and dry. The lack of nutrients and moisture makes them frail, causing them to break off easily.

How to Determine the Appropriate Watering Frequency

There are several factors to consider when calculating the appropriate watering frequency for your specific succulents and cacti, such as size, plant species, growth stage, soil characteristics, environmental conditions, etc. To determine the appropriate watering frequency, follow the following guidelines:

1. Understand the Species

Succulents and cacti have different sub-species, each with varying water needs. Some are more drought-tolerant, while others require more regular watering. Research the specific water requirements of your cacti and succulent species.

2. Consider Growth Age

The growth age of your plant must be taken into account when watering. If your plant is young, it will require frequent watering, and if it is old, it won't need as much water as the young plant.

3. Evaluate Environmental Conditions

Plants' water requirements and frequency are influenced by their exposure to heat and direct sunlight. Cactus plants require exposure to bright light and warm temperatures, which increases the water evaporation rate. So, ensure your watering schedule is adjusted according to your plant's environment.

4. Check Soil and Potting Mix

Use well-draining soil or a specialized cactus mix to prevent waterlogging and root rot. The soil should allow excess water to drain away quickly.

5. Learn the Signs of Thirst

Pay attention to your cacti and succulents, and watch out for signs of dehydration. Some signs of dehydration include slight wilting, dried spines, root discoloration, or shrinking of the plant. When you notice these signs, water your plants.

6. Perform the Finger Test

The finger test is a cool trick to learn as a seasoned gardener of succulents and cacti. Stick your finger into the soil about an inch (2.5 cm). Pull your finger back up after reaching that depth, and check for signs of moisture. Water only when the soil is completely dry at that depth.

7. Use the Soak and Dry Method

When you water your cacti and succulents, water them thoroughly until the water drains out of the bottom of the pot. Ensure you allow the soil to dry completely before watering again. Avoid frequent light watering, as this can lead to shallow root growth.

8. Water Sparingly in Winter

During the dormant winter months, cacti and succulents require less water. Reduce watering frequency to 30-50 days to prevent overwatering during this period.

9. Avoid Watering the Crown

Water the soil around the base of the plant while avoiding direct contact with the plant's crown (the central part from which the spines emerge). Excessive watering of the crown can lead to the plant rotting.

10. Use a Watering Schedule

Develop a watering schedule based on the specific needs of your cacti and succulents. Keep records to help you keep track and adjust when necessary.

11. Controlled Irrigation

An automated irrigation system provides water at specific intervals, ensuring consistent moisture levels.

How to Choose the Right Watering Tools

When watering cacti and succulents, it's essential to choose the right tools to provide water in a controlled and precise manner. Shown below is a selection of ideal tools for you to choose from.

- **Watering Can with Narrow Spout**

Choose a watering can that has a long and narrow spout. This type helps you target water directly at the base of the plants. It also helps you avoid getting water on the plant's body or crown, preventing rot or other problems. A watering can with a small capacity is perfect for cacti and succulents, which generally prefer infrequent but thorough watering.

- **Spray Bottle with Fine Mist Setting**

Choose a spray bottle with a fine mist setting that will gently water the small cacti and succulents. It is handy when you're propagating succulents or watering baby plants with delicate foliage. This bottle can also help you avoid overwatering because the mist controls the amount of moisture the plants get.

- **Watering Wand with Adjustable Flow**

Another good piece of equipment to own is the watering wand. This instrument has an adjustable flow for larger cacti, succulents, or multiple plants in the same area. The adjustable flow allows you to control the water pressure, providing a steady stream to reach the entire root zone.

- **Tray and Saucer**

Trays and saucers are necessary for the bottom watering method. This method is a good option for cacti and succulents with sensitive leaves or densely packed spines. It's done by placing the potted plants in a tray or saucer filled with water and letting them absorb water from the bottom up through drainage holes until the topsoil is moist. Empty the tray after some time.

- **Use Well-Draining Soil and Containers**

Use well-draining soil and containers with drainage holes to prevent water from sitting around the roots too long.

How to Assess Soil Moisture Levels Effectively

Knowing a plant's soil moisture is essential for determining when to irrigate. However, it can be challenging to work out when succulents and cacti have accurate soil moisture. You can assess soil moisture properly by using the techniques listed below:

- **Use a Moisture Meter**

Place the moisture meter probe into the soil at different depths to measure moisture levels. This meter will indicate whether the soil is dry, moist, or saturated.

- **Perform the Finger Test**

A finger test is another good way to check for the moisture content of the soil. If the soil feels dry, it needs watering. If it feels slightly moist, it's adequately watered. Saturated soil will feel overly wet.

- **Check the Weight of the Container**

Lift up the container to gauge its weight when watering the plant. Lift it regularly to feel the difference in weight when it becomes noticeably lighter (Wet soil will be heavier than dried soil).

Water Quality and Temperature Considerations

The quality and temperature of the water are both things that need to be checked before you use the nearest water outlet. Not all water is suitable. Here are some factors to monitor.

- **Effects of Chlorine**

Chlorine, a common disinfectant found in tap water, can have detrimental effects on the delicate balance within your cacti and succulents' root systems. It can destroy organisms necessary for plants, thereby damaging plants' roots. Therefore, do not use chlorinated tap water on your succulents and cacti.

- **Effects of Fluorine**

Be on the lookout for fluoride in your water as well. Excessive fluoride levels can be toxic to your plants. Be cautious when you use tap water, not just because of chlorine alone but also for fluoride, as some regions have naturally high fluoride levels. Using filtered water can significantly reduce the risk of fluoride toxicity in your cacti and succulents, promoting their longevity and vibrancy.

- **Excessive Mineral Content**

Water rich in minerals gradually leads to the accumulation of salts in your cacti and succulent soil. This can potentially harm the root systems of your plants. Regular soil flushing is a good way to prevent this build-up. Every few months, water your plants thoroughly, allowing water to flow through the soil and wash away excess salts. This practice will help healthy nutrient absorption and root growth.

Congratulations! With this knowledge, you are well on your way to owning an amazing garden with the best-grown succulents. When next someone talks about a succulent-cacti myth, you'll be able to correct it. Every garden needs succulents and cacti, and not just any kind. Well-grown ones will boost your garden's aesthetics. With your newfound understanding, bring that dream garden to life.

Chapter 5: Cacti and Succulent Health, Care, and Maintenance

Adopting a cactus into your home is usually an effortless process and adds a snap of contrast and color. As they come in almost all shapes, sizes, and colors, blending them in with your decor should not be a massive challenge. Whether you choose to add them to your home, plant them in your garden, or place them on your windowsill, it is crucial to be acquainted with your new house guest's needs. Cacti are known for being tolerant and low-maintenance plants that require almost little to no care. However, a common misconception is that all cacti grow in the desert. That little tidbit is not true. Some cacti actually grow in the forests and are used to living in humid and moist environments. Some cacti can withstand the most arid environments, from the hottest weather to below-freezing cold temperatures.

To be able to properly care for your trouble-free house plant, your best bet is to recreate its natural habitat. Desert cacti can thrive indoors. With the proper care and proactive maintenance, you will be left with a beautiful symphony of colors and unusual formations that please the eye and, in some cases, have healing properties.

Desert cacti thrive indoors.
https://www.pexels.com/photo/the-sun-shining-through-the-blinds-into-a-room-with-a-sofa-and-a-plant-18246654/

Catching any irregularities in your plant early on prevents complications from becoming a problem. For example, making sure there are no pests or rot in the early stages of growth can prevent problems from spreading to the rest of the plant. In some cases, early diagnosis can protect neighboring plants if it's a tight formation of several cacti or succulents arranged together, whether in a garden or an indoor pot.

As you go through this chapter, you should be able to identify the signs of a healthy cactus versus one that may be struggling to sustain itself. You will find essential information on how to care for and protect your succulent from withering.

Essential Care Practices

There are some basic practices to get you started on the right path to owning healthy and vibrant cacti. Some of these points you may have missed during the initial cactus care. Here are the first things you should check off your list for a hassle-free cacti care routine.

Let There Be Light

After verifying the type of cactus you have and the natural habitat you need to create for it to flourish, consider the amount of light you're exposing it to. Most cacti enjoy bathing in the sun, even indoor plants. Find a south-facing window or corner in your garden to place your cactus. Make sure you don't put it in the direct, hot sun, as that can result in a yellow or brown colored cactus (sun-burned) instead of vibrant green or colorful hues.

If your cactus does turn yellow, try moving it to a place shaded from the light to cool down. Also, don't put it near air conditioning or a strong draft since dry and warm is their love language.

For forest dwelling cacti such as Rhipsalis, you need to provide shade to protect it from direct sunlight, whereas its cousin Echeveria thoroughly enjoys its days in the sun.

In the colder months of the year, around autumn and winter, it's essential to maintain a cool climate of around 46.4°F to 50°F. As the summer and spring roll around, the plants need more air but will be just fine exposed to high temperatures.

Indoor cacti are also known to display quicker growth patterns when placed out during summer nights when the weather is around 122°F.

Make Sure They Are Adequately Hydrated

Cacti are more like camels. They store water for long periods and don't need that much in the first place to survive. There is a fine line between overwatering and underwatering. Each situation has its drawbacks. If you add too much water to your plant, you're delaying and impeding its growth. It can also rot the roots, and your plant could develop scabs that look like oxidized areas on the stem. If you add too little, your cactus will shrivel, wither, and eventually die. Obviously, you don't want that for your greenhouse guest.

A way to check if your cactus is thirsty is by checking the soil. Have a look at the top two to three inches, and if it is completely dry, it's time for a drink of water.

When it comes to watering, if you can, and you have easy access, avoid tap water and use a more natural source, such as rain. That's because tap water contains certain minerals that collect in the soil and are detrimental to your plants. They can also throw the flow of nutrients that the plant relies on into disarray.

Watering during the hot months should be done around once a week. In between watering, you need to make sure that the soil is draining properly and allow it to dry a bit before the next time. If there is excess water that hasn't drained, make sure you empty it.

Watering is a bit more subjective in the colder months, depending on the environment and how fast the soil dries out between watering times. This is a time when the plants enter what is called a resting period. For cacti that flower during the winter, make sure you're providing a warm environment and regular watering. For desert dwellers, you don't need as much water as you think, and they can be left alone for extended periods.

Maintain the Soil

Making sure that the soil drains properly is step one. Making sure your soil is aerated is step two, and maintaining the correct mix of soil, sand, and organic material is the third step towards healthy, happy cacti.

Follow the instructions for your fertilizer to the tee. According to the printed directions, some cacti nutrients need to be diluted before use. As with watering, do not over or under-feed your plant.

Clean Your Plant and Its Surrounding Environment

Every few weeks, take a damp cloth to your cactus and wipe it down to make sure you get rid of any dust or debris stuck to it. Make sure you use protective gear and gloves so you don't come away with a bloody hand.

While most cactus pruning is done to propagate new plants with tiny pups, it is also considered a part of cleaning. Pruning the cactus can be done as a method to control its growth or remove the dead and damaged bits before they spread. This is done using clean and sharp garden shears. As the cactus starts growing, rearrange the rocks you initially added for warmth and protection.

Prevention Strategies for Pests and Diseases

A deterrent to pests and diseases in cacti can come from simply being proactive. Do not wait for your cacti to show signs of ailments that may lead to cutting parts of them off or using harsh chemicals to try to treat them. There are a few techniques that can be followed to prevent pests and diseases.

Inspecting the Roots

Take a close look at the roots of your cacti every two to three months. Some diseases may take months to show physical signs on the plant. You can limit your plant's ordeal by checking its roots, as most problems start from there and work their way up the stem and leaves.

Overwatering

Overwatering your cacti can easily lead to rotting roots. Adding too much water or watering more frequently than needed can be a fertile ground for fungal diseases to grow. So, instead of reaching a point where you need to trim the roots and repot the plant in the hope of saving it, follow the appropriate instructions to water your plant, making sure the soil drains properly.

Maintaining Good Hygiene for Your Plant

Healthy plants are a lot less likely to attract pests and diseases. Ensure you carefully follow the care instructions specific to your cacti species. This includes proper sunlight without overexposure, proper watering techniques, and an adequate soil mix specific to cacti that drains properly. Also, make sure you periodically clean out the pot from fallen leaves and dead flowers.

Quarantine Your New Additions

To avoid the transfer of pests and diseases that may have gone unnoticed before you bought your plants, make sure you separate your new plants from the old ones. Inspect the newcomer's roots and leaves carefully until you're satisfied they are pest and disease-free. If you discover a problem, treat it immediately or dispose of the plant to avoid contaminating the rest.

Make sure you separate your new plants from the old ones.
https://pixabay.com/photos/fat-plants-terracotta-pots-2734948/

Insecticidal Soap

Before you apply this to clean your plant, make sure you read the label very carefully, as some of these are harmful to plants. Check and double-check to make sure that they don't have any negative side effects. Even after you do, only use it on a small part to test it before spreading it over the remaining parts.

Underwatering

Even though you opted for a resilient plant that needs very little water every now and then, it still does need water. Forgetting to hydrate the cacti is a sure way for it to wither and die from the roots up.

Use Organic Fertilizer

Non-organic fertilizers can also be a source of pests and disease. They contain varying amounts of heavy metals like zinc and copper that are harmful to the plants. These fertilizers weaken the plant and leave it vulnerable if it's attacked by pests and diseases. Choose organic fertilizers and foods as much as possible.

Physical Trauma

Treat your cactus gently. When moving it around, potting, and repotting, make sure you don't bump it into any sharp or heavy objects. These physical injuries make it more susceptible to infections.

Be Mindful of the Temperature

Temperature fluctuations can easily affect your cacti, especially if your area doesn't resemble its natural habitat. For example, try moving the cacti inside your home during cold winters to keep it warm and away from the harsh cold. Humidity in the weather can also cause disease, depending on your cacti species. While some thrive in humid weather and grow faster, others may prefer a drier environment to stay clear of fungi and pests.

Soil Mix

As previously mentioned, cacti benefit from a mix of different ratios of soil, sand, and organic material in its soil. However, increase one or the other rather than using a similar mix. For instance, too much sand can easily harm your cactus, making it hard for it to absorb water and nutrients from the soil. Be mindful of the proper mix for your plant prior to planting.

Nutritional Needs and Fertilization

There is more than one option to choose from when it comes to the fertilizer most suitable and safe for the species you're breeding. Whether it is organic or industrial, the first and foremost rule is to follow the instructions. Don't over or under-feed. Don't use fertilizer that could harm the plant. And do not experiment without knowing what you're doing.

The reason you should consider extra nutrients for your cacti is because of the soil composition used for the plant. The mix of sand and easy-draining medium doesn't promote nutrition retention enough, and it's easily washed away before the cactus can benefit from any of it. So, an extra helping hand is needed.

Below, you'll explore the most common and recommended nutritional additives that can benefit your house or garden plants.

Primary and Macronutrients

There are three main macro and primary nutrients that every plant needs to thrive and grow. Nitrogen promotes a healthy stem and foliage, phosphorus for strong blooming and steady, healthy roots, and potassium to assist in absorbing the needed nutrients and to fight off disease.

As you select the fertilizer, you will notice that all of them have the ratio of N-P-K stated on the back of the package. Now, as important as it is to get the right fit and ratio for your plant species, a commonly recommended ratio of the N-P-K is 6-6-6, which usually supports the growth of the cactus.

There are other types where you may find the ratio ranging from 10-10-10 to 20-20-20, which is also not a bad choice, but it may need to be watered down a bit more than the others to avoid over-fertilization and the risk of damaging the roots.

Picking out a fertilizer with added microorganisms improves the conditions in the pot to compensate for the low nutrient levels.

Organic Fertilizer

This is your best choice for fertilizing your cacti. There are quite a few to choose from, depending on your needs.

Compost Manure – If you live near a farm and are growing outdoor cacti, natural animal manure is known for retaining and providing priceless nutrients that benefit your plants.

Whether it's from sheep, cows, or chickens, any of these options should provide you with a green and environment-friendly fertilizer for your plants.

You can use it indoors as well, as long as you keep in mind that the smell may linger a bit longer in a closed space.

Worm Castings – These are basically the manure of worms. This type of fertilizer is rich in humus, improves the ventilation of the cacti soil mix, and balances the pH levels. It is rich in over 60 micronutrients. Along with the N-P-K, they contain amounts of magnesium, carbon, iron, zinc, and many other nutrients. They also assist with absorbing heavy metals from your soil mix, lowering the chance of your plant retaining them in toxic amounts.

Tea Bags – These are little manure tea bags that you soak in water and add to your plants. It comes in the form of biodegradable tea bags. To use these, you need to soak them in about five gallons of water somewhere between 24 and 36 hours until the water turns a golden brown. You then use the water the next time you need to hydrate your plant.

Liquid and Granular Fertilizer

Fertilizers come in many forms: powder, liquid, or granules. Liquid and powder fertilizers are diluted before use on the cacti. The roots easily absorb these nutrients but are unfortunately also washed away easily, so more than one application may be needed for an optimal result.

As for granular fertilizers, they are broken up and spread over the soil mix and then watered in. These types are more concentrated and last for up to nine months. The resulting improvement can be witnessed within two weeks. Make sure that the N-P-K ratio is low to avoid root burn.

Fertilizer Spikes

These, like granules, are considered slow-release fertilizers. They don't make a mess and are hassle-free to use. All you need to do is place the spike in the potting mix at the start of the growth season and water the plant as needed for that species. They also last for nine months and, as they are not visible, they are child and pet-friendly.

Re-Potting and Transplanting

As your plant grows in size or if you wish to relocate a newly bought one, you're probably going to consider re-potting in a new pot. Another reason you may consider this is if the plant was infected with pests or diseases, and you need a clean and fresh reallocation while disinfecting and cleaning its old home. There are a couple of straightforward techniques to take into account to maintain a healthy cactus and to perform the repotting successfully.

A pot with holes in its base to drain the excess water is the best option. Good drainage protects the cactus roots from rotting. However, that doesn't completely rule out the idea of using a regular pot; you just need to take a little bit more care when watering it.

Always make sure that the soil is dry all the way down before watering it, and use ¼ cup or ½ cup of water every week or two. This should allow your cactus to be content.

Now for the actual procedure of a healthy repotting;

- Make sure your plant is watered well before the procedure and properly drained from any excess.
- As you remove the cactus, use protective gloves and folded paper towels to protect your hands.
- Find a way into the old soil using chopsticks to separate it from the roots without damaging it.
- Add the potting mix to the new home and make sure it's slightly wider in diameter, situating your plant on top.
- Add more potting mix and firm down the soil with your hand.
- Keep away from watering for a few days to prevent damaging the roots.

Seasonal Care Adjustments

Caring for your prickly friends differs slightly from one season to another. There are a few differences in watering and placement that should be taken into consideration as colder or warmer weather comes around.

Spring and Summer

As the weather warms up, you should reconsider your watering schedule. If you are in a sunny environment where your plants may be subject to a lot of heat, it is advised to water the cacti once a week. Before watering, make sure the soil is well-drained and dry. Be careful not to burn your cactus by pacing it away from direct sunlight. Place it in a shaded area with enough access to the sun without causing the leaves and stems any discoloring. Also, make sure that the north side of the cactus is facing north, which should be marked beforehand. If you're keeping your cacti indoors, it would be preferable to use a south-facing window. Light is not limited to the sun only, as you can keep it in bright light using indoor lamps.

Autumn and Winter

As the colder months roll around, you may need to cut back on the water a bit. As always, make sure the potting mix has dried out completely between watering. The frequency you choose to water your cactus depends on the weather and environment they are in, along with how different the place they are in is from their natural habitat. Desert cacti and cacti that flower in the winter need different care. While one may thrive on little water and resist the brittle cold, the other may need warmth and extra watering sessions.

Make sure your cacti are well protected from harsh winter drafts, especially if they are indoor cacti or aren't desert dwellers accustomed to colder nights.

Around late February, close to the end of winter, consider sprinkling "Preen," a weed preventer that makes the weeding task much less arduous.

Chapter 6: Pruning and Shaping Stunning Succulents

Succulents require routine upkeep because they grow slowly and eventually outgrow their containers. If you notice that your once neatly cultivated succulents look unkempt, it's time to get out the pruning shears.

Pruning and shaping succulents is not a complex matter. You need to know just a few details to make the process smooth. This chapter will guide you on what this process is, what the benefits are, and the tools and safety precautions needed for the work. Furthermore, you will learn some techniques and how to manage overgrowth and pruning for succulent health and beauty, which, in the long run, will extend the life of your plants.

Pruning and Shaping

Pruning is removing or cutting off a part of the succulent plant, vine, or tree that is not needed for growth or is dangerous to the overall health and development of the plant. Conversely, shaping is a specialized form of trimming focused on designing and maintaining the unique form of the succulent plant. Its application is seen especially in topiary, pollard, etc. In horticulture, pruning, and shaping maintain and boost the plants' growth pattern, size, health, and overall appearance. This is carried out to enhance your plant's growth, and it is an efficient way to maintain developing and established succulent plants. It also occasionally helps protect people, properties, and plants from disease, pests, or damage.

This is a long-term maintenance technique.

Pruning is removing or cutting off a part of the succulent plant, vine, or tree that is not needed for growth or is dangerous to the overall health and development of the plant.
https://www.pexels.com/photo/a-plant-on-a-clay-pot-9414306/

Benefits of Pruning and Shaping

Promoting Branching: Pruning and shaping to remove or cut off any stagnant branch encourages new growth, leading to a more robust and beautiful succulent plant.

Maintain Plant Health and Aesthetic: When you remove dead, decayed, or diseased parts of succulent plants, it stimulates the aesthetic appeal and the overall health and appearance of the plant. The removal of the dead leaves closes the door to any and every secondary attack of pests.

Controls Growth: Growth control entails pruning the overall size and density of the plant. Shaping, in this case, is used to create a shape for your plant that is pleasing to the eye as well as good for the plant.

Create Distinct Forms: This technique enables you to create unique, specialized designs like topiaries, espaliers, pollards, and hedges.

Encouraging Symmetry: Creating uniformity and enhancing symmetrical appearance through this technique increases the overall visual appeal of the succulent.

Encourage Flower and Fruit Production: Pruning and shaping succulent plants, shrubs, and trees improves the production of fruit and flowers. This is done to stimulate an open canopy to penetrate more light, encouraging the production of a flower bud.

Tools and Safety Precautions

Apart from maintaining the beauty of your landscape, pruning and shaping is vital to promote the health and growth of your succulent plants. The right tools are necessary to get the best results. Below is a list of must-haves, essential tools, and equipment used for this task:

Pruning Shears

Pruning shears are the most frequently used tools for shaping a tree's flowers, shrubs, vines, and minor growths. Pruning shears are designed to be hand-held and can cut up to ¾ in sticks and tree branches. Pruning shears come in three types: ratchet, anvil, and bypass.

Ratchet pruners are like the anvil but were designed to cut in stages. It is a good fit, especially when you don't want to stress your wrist. The anvil pruner has a straight blade that can split dry branches and stems. A bypass pruner is the most well-known among the three and functions like scissors.

Loppers

Loppers are needed to prune and shape vines, nuts, and fruit trees. They can cut branches to the range of 2 1/2 inches thick. Loppers are similar to hand shears, but the blade is thicker, while the handle is very long. Like the pruner, loppers also come in three styles.

Pruning Saw

A pruning saw comes in different sizes and is great for pruning branches ranging from 1 to ¾ in diameter. Folding the pruning saw makes carrying it around your farm or garden easy. The pruning saw is different from the handsaw or hacksaw you see in department stores. It has a curved, narrower, and pointed blade that can easily remove branches when used in a dense environment. The teeth on the pruning saw are well-placed, which causes the saw to cut either way, whether you are pulling or pushing.

Pole Saw and Pole Pruner

You know those small tree branches that are challenging to reach? Well, a pole saw can reach them and cut them. It's safer to prune from the ground instead of using a ladder, and a pole saw gives you that leverage to prune and shape safely from the ground. A pole saw is a saw blade joined to a long pole. A pole pruner, on the other hand, comprises a fixed hook and hung blade controlled by a rope and attached to a long wooden pole. You can cut branches up to 2 inches in diameter using a pole saw and pole pruner.

Thorn-proof Leather Gloves

Some succulents can literally be a thorn in your hand, no matter how pretty or harmless they might seem. Nonetheless, you don't have to dress like you're going to a beehive to get some honey to tend your succulent plants. Scratches and cuts from succulent plants with sharp edges can be prevented by using thorn-proof leather gloves that protect you up to your elbow from being hurt by thorns.

Protective Goggles

These goggles might be big, looking all-nerdy, but on the bright side, they are the very thing that protects your eyes from spider webs, insects, and other foreign bodies. They also protect against poisonous chemicals that get released into the air and nasty allergenic plant pollen. These dangerous substances could lead to problems when they come in contact with your eyes, hence a need for protective goggles.

Step-by-Step Pruning Techniques

Pruning is a vital part of ensuring the health of your plants, and a proper technique is a plus to accomplish this. Here are some step-by-step succulent pruning techniques.

Tip Pruning

Tip pruning, also known as pinch pruning, is the removal of the end of each shoot during the growing season, either with a thumb or a finger on less woody shrubs. At every point where a cut is made, tip pruning nurtures the growth of more shoots at that cut point, leading to a more rounded, succulent plant with more flowering stems. Furthermore, tip pruning gives succulent plants a round, light trim to restore their shape after flowering. To accomplish tip pruning, especially on larger jobs like topiary, use sharp shears or hedge trimmers to give you distinct lines.

Stem Pruning

Stem pruning removes or cuts off distinct branches or stems from the succulent plant. This technique is also used on some herbaceous plants, shrubs, and trees. When and how you use this technique is determined by factors like your desired outcome, the species of plant, and their growth pattern.

Selective Leaf Removal

As the name implies, selective leaf removal is the practice of removing leaves from the stem of succulent plants, especially around areas with clusters of fruit. Selective leaf removal on a succulent encourages air circulation, exposes fruit to more sunlight, enhances insecticide penetration, and reduces herbaceous pH and potassium. Furthermore, it enhances the plant's bud fertility and color. While using the selective leaf removal technique, you must leave some leaves on the stem to encourage the production of more carbohydrates to aid fruit development, growth, and overwintering reserve development. Removing three to four leaves is often enough to achieve this technique.

Shaping Methods

Shaping methods are ways of making definite plant designs. The focus of shaping your succulent plants is to show their beauty and enhance the landscape, allowing farmers or gardeners a little creativity and artistry here and there. Below are some different shaping methods.

Topiary

Topiary is the act of training and pruning your succulent plants, trees, and shrubs into a decorative shape. It is a method where you diligently trim the succulent plant into a pyramid, ship, animal, cone, or any form of imaginative design you choose. Topiary shapes are a good way of adding unique shapes to your garden. It doesn't matter if they are found in a steel container outside your house or among perennials in an enclosed environment. They are the statement plants of your garden.

Topiary is the act of training and pruning your succulent plants, trees, and shrubs into a decorative shape.
https://unsplash.com/photos/green-trees-under-white-sky-during-daytime-QeVdVJ3_2sg

Here is how to shape the topiary.

- To shape a ball, get a length of garden wire and construct it into a circular shape that you can hold easily. Then, move it over the succulent plant you want to shape. Make the sphere frame smaller than the foliage's mass if you want a perfect shape.

- To restore a cone shape, stand over the plant and start shaping from the center, working your way outwardly around the plant. As you work, you can rest a cane on each side of the plant to form the shape as a guide for cutting. Fasten the canes at the top to form a wigwam shape, and with a garden wire, bind the sides together. Use your shears to shape the plant within this framework. Using this technique, use a handheld trimmer that looks like sheep shears to get maximum results. It will give your work a clean finish.

Espalier

An espalier method is used to train your succulent plants, especially the ones with fruit, to rest against a support such as a fence or a wall, with various tiers of horizontal branches and a central vertical stem. To create these tiers, you will have to do your first pruning soon after planting and continue to do it annually in summer to keep the plant and its fruit in good shape. The Espalier method is mainly used to train plants to grow against walls or in smaller gardens to maximize space.

An espalier method is used to train your succulent plants, especially the ones with fruit, to rest against a support such as a fence or a wall.
https://pixabay.com/photos/espalier-facade-building-decoration-5032964/

Bonsai

Bonsai shaping is the method of growing and training your succulent plants, shrubs, and trees in a box or container until they become full-sized, mature trees in a mini form. This practice is from China, and its focus is to create a well-balanced, appealing design of trees in a limited area of space. A bonsai must be pruned as often as possible, and it involves two types of pruning: maintenance bonsai pruning and structural pruning. These two processes are vital to bring the bonsai into the desired shape and keep this shape for a long time.

Bonsai shaping is the method of growing and training your succulent plants, shrubs, and trees in a box or container until they become full-sized, mature trees in a mini form.
https://pixabay.com/photos/bonsai-azaleas-rhododendron-3125722/

- **Structural Bonsai Pruning:** Using pruning, you can create a beautifully shaped bonsai succulent. Unlike the maintenance cut of bonsai, structural pruning is more radical and requires you to be well-grounded and prepared. Structural trimming usually occurs at the bonsai's early developmental stage.
- **Maintenance Bonsai Pruning:** This technique aims to maintain the definite style of the plant while promoting its existing shapes in small, easy-to-achieve steps.

Managing Overgrowth and Legginess

Succulent plants need to be pruned every few years. They become overgrown and leggy when not properly tended. So, how do you manage overgrowth and legginess?

Strategies

- **Take Stock of Your Plants:** You will suddenly realize how well some plants are doing compared to others struggling with overgrowth. Take note of your plants and their current state. Move plants around so that each plant has the best growing conditions you can provide.

Causes

- **Normal Growth Pattern:** The plant's growing conditions affect it both positively and negatively. When your plants are supplied with enough nutrients, water, soil, and sunlight, they will grow normally and healthily. On the other hand, when the growing condition becomes 100% perfect, there is bound to be an overproduction of foliage, so much so that the young roots and stems cannot support it. It gives rise to leggy growth and weak plants. This can be frustrating for a gardener who has invested time and effort. No gardener wants to see their cactus plant looking sick.

- **Inadequate Light:** When exposed to low light, plants cannot photosynthesize and make the glucose required for their growth. Hence, the leaves and stems of the plant move upward in search of light. This stretching is what is known as leggy growth.

- **Rapid Growth:** The high nitrogen content in some fertilizers causes vibrant plant growth. Nonetheless, excess nitrogen in the fertilizer can cause rapid, tall, and skinny-looking plants, producing poor flower and fruit yields. Furthermore, rapid growth makes your plant top-heavy, unable to support itself, and more prone to disease.

- **Lack of Pruning:** Plants need pruning at one point or another to remove too many dead leaves and leggy growth. If you want to foster thicker, sturdier branches and stems, it is best to prune your plants early during spring. Your plant's health will diminish if leggy stems are neglected.

- **Nutrient Imbalance:** A nutrient deficiency in your plants leads to poor flower and fruit production and leggy growth. This is why regular fertilizing is a must. Nonetheless, when you fertilize plants too much, it causes leggy growth. The way out is to use a diluted water-soluble fertilizer. Furthermore, stop fertilizing regularly if you are still seeing leggy growth. Let the plant recover by itself.

Corrective Measure

- Put the plant in a deeper container to allow the stem to be completely immersed in the soil. Plants that have hairs on their stem can grow into roots.

- When planting in your garden, ensure the stem is buried completely in the soil. The soil serves as a source of support and

provides extra nutrients for the stem.
- Expose your plants to direct sunlight or ensure you have a grow light to enable your plants to photosynthesize easily and make enough food needed for healthy growth.
- Use a good watering system and ensure the soil is drained properly.
- Do not over-fertilize. This is usually the cause of legginess.
- Prune regularly to remove leggy foliage and dead leaves. This will stimulate airflow to lower leaves and prevent plants from giving nutrients to unhealthy foliage.

Pruning for Succulent Health and Beauty

Techniques for Improved Air Circulation

Allowing air to circulate is vital for the overall health and appearance of your succulent plants, and here is how to achieve this.

- **Removal of Dead Leaves**: Remove dead or decaying leaves as you prune. Don't leave them around your plants where they could become home to pests and too much water.
- **Prune Overgrowth and Legginess:** Thin out areas in your garden that are overgrown by removing some leaves, stems, etc.
- **Lift the Pot or Container off the Ground:** This technique allows air to circulate around the pot as you lift it from the ground.
- **Place Your Pots in an Airy Location:** Look for areas with enough air and place your succulent pot there. This will help minimize the risk of damp soil that gives rise to diseases.
- **Avoid Overwatering and Over-Fertilizing:** Over-fertilizing and watering results in leggy growth. Water your plant only when you notice that the soil is dry.

Identifying Signs of Unhealthy Growth or Damage

- **Leaning Towards the Light:** When your succulent plants begin to bend towards the light or lean toward the direction of your window, it means that your plant is in serious need of sunlight. Move your plant toward a source of light to correct this.
- **Shriveled Leaves:** Dying leaves are normal in the succulent plant world. You may even start pruning your succulent plants by taking out some leaves when they look unkempt. However, when

all the leaves begin shrinking, your plant needs water. Firstly, see if the soil is dry. If yes, water your plant, ensuring the soil is well-drenched. Furthermore, keep a schedule for watering your succulent plants.

- **Rotting:** You will know you are experiencing rot when the base of your plant and its leaves get mushy. This is due to too much watering, with the soil becoming waterlogged. Wait for the soil to dry out if you sense that the rot is caused by over-watering. If the rot persists, then you need to consider trimming your succulent.
- **Yellow Leaves**: Yellow leaves are caused by over-watering. Refrain from watering until the soil becomes dry and the leaves start regaining their color.
- **Black or Brown Spots**: If your plant turns black or brown along the trunk or at a certain spot, your watering is slightly over the top. Refrain from watering and observe the trunk, as chances are that the spot will disappear. However, restoring such plants will be difficult if the trunk has turned completely black.
- **Dull Color**: This is caused by too much exposure to sunlight, which results in bleaching. You can see a purple or yellow plant turn lighter green or a green succulent turn white or pale green. Moving your plant away from sunlight to a less-bright corner can help in this situation.

Pruning and shaping is one of the most daunting and misconstrued landscape maintenance strategies for homeowners, farmers, and gardeners, and most individuals don't know what to do or how to do it. Understanding what this technique entails and which tools to use should clear up some confusion and fears you may have had on how to manage the overgrowth of your succulents. Take your time to observe your plant and act accordingly.

Chapter 7: Propagation Techniques and Maximizing Yield

Grown, nurtured, and pruned your very first succulent species to your heart's desire? Great work! Do you wish you could collect more of the same plant, down to its genetic makeup? Your wish will be granted in this section. Welcome to the fabulous and intriguing world of propagating cacti and succulents, where the science of gardening meets the magic of creation.

Propagation is the reproduction of plants. As a gardener, aspiring or experienced, you may already know that plants generally reproduce through pollination. It is the process of merging two parts of different flowers – pollen grains of the male with the stigma (central part) of the female – to produce the offspring (seeds).

Propagation is the reproduction of plants.
Propagation at Peckover by Barbara Carr, CC BY-SA 2.0 <https://creativecommons.org/licenses/by-sa/2.0>, via Wikimedia Commons: https://commons.wikimedia.org/wiki/File:Propagation_at_Peckover_-_geograph.org.uk_-_3610104.jpg

It's like a platonic variant of human reproduction in the sense that the two flowers don't mate together. Pollination occurs when an external agent (insects, wind, water, etc.) collects the grains from one flower and drops them in the stigma of the other.

If you have put your thinking cap on, you may have already started imagining how you could enact the role of an external agent to breed your very own batch of succulents. Hold your horses, though, and think some more. Pollination will give you a hybrid, not the same plant with identical DNA.

To duplicate your favorite cactus's genetic makeup, you will need to learn the technique of cloning (yes, it is real, not sci-fi!). To understand that technique better, you will need to ascertain why you should propagate your plants in the first place.

Benefits of Propagation

The propagation of cacti and succulents has a number of personal benefits as well as various ecological advantages. It has a positive impact not only on humans as a species but also on the environment as a whole.

- **Expanding Your Collection**

How many times have you succeeded in growing succulents from seeds? And how many times have you ended up buying a brand new, full-grown plant instead? The next time you wish to expand your succulent collection, you don't need to spend any more money on buying seeds or plants. With propagation, you can fill your backyard with as many cacti and succulents as you want.

- **Choose What You Grow**

Do you want a particular species of cactus or a specific type of hybrid? Plant shops and nurseries usually don't stock every species, and even if they do, they won't have more than one or two on sale. Once you know the art of propagation, you won't ever have to search for your favorite hybrid. Simply grow one from your existing plants, multiply your collection of a particular species, or experiment to develop a new one! Choose exactly the kind of cacti and succulents you want.

- **Share with Others**

Say that you and your neighbor have been growing cacti and succulents for so long that you each have your own comprehensive collection. But you don't both have the same varieties. What if you work together and share your collections with each other? Propagation allows you to do just that. While sharing, you don't incur any losses. You just need to hand over some small part of your plant to them, which eventually grows back.

- **Save Money**

Many kinds of cacti and succulents are priced in the lower bracket. But while expanding your collection, those costs will add up. Propagation is much cheaper and, in certain instances, free. You can save a ton of money in the long run after learning the art.

- **Prevent Extinction**

Quite a few species of succulents are rare, and a few others are hard to grow. Some of them even require very specific conditions to thrive that aren't easily found. All these factors combined may result in the extinction of species like the astrophytum asterias or star cactus. Propagation can prevent their extinction by multiplying them, which is slightly easier than growing them from scratch.

Mastering Propagation Techniques

The benefits of propagation are appealing and may have prompted you to try out this art for yourself, especially if you have grown up watching sci-fi movies like the Star Wars franchise. But before you learn how to clone your cacti and succulents, it's important to understand and master the process of seed propagation.

Seed Propagation

A Fair Warning: It is much easier and faster to buy new seeds than to pollinate them from existing cacti. But suppose you want to create a special hybrid, grow succulents without spending any money, or simply experience the joy of creation. In that case, seed *propagation is the way to go.*

1. Wait for your succulents to mature. It may take anywhere from a year to over 50 years. Once they start blooming, you can begin the process of propagation.
2. It is better to keep the two plants that you're about to pollinate beside each other. For instance, the best way to pollinate a bunny ears flower is to shake the male flower directly over the female stigma of the other plant.
3. Alternatively, you can catch the pollen of the male with a paintbrush. Dip the brush in rubbing alcohol to clean it thoroughly. Rub it over the male flower's center, where you will notice white dust-like dots. That's pollen. Let it get smeared around the entire brush.
4. Bring the brush to the female flower's center, where the stigma is located. Ideally, you will just need to push the brush in lightly to stick the pollen to the stigma. But if it doesn't stick, then it could be that the stigma isn't open yet. Wait for a few days until it opens and the particles stick.

Once the pollen has mated with the stigma, all you need to do is wait. The process may produce seeds within a month, or it may take an entire year. Then, simply collect the seeds and plant them in a different pot in the same way as described in Chapter 3. You will be able to grow the offspring of the two plants, which may be the same kind of plant or an entirely different hybrid, depending on which seeds you have managed to propagate.

As you may have noticed, seed propagation may take around a year or more to begin developing a new succulent. The cloning process, on the other hand, is much faster and easier.

Stem Cuttings

As the name suggests, cut stems of your cacti and succulents are used to grow new plants. This method has a two-way benefit. You can prune the old plant while growing a new one. Succulent stems grow larger and longer than their leaves to absorb more sunlight. When they reach the point where your ornamental plant starts looking weird and out of place, you can cut the stems and wait for a new one to grow.

But don't throw the old stem cuttings away. You can use them to propagate more of the same kind of cactus or succulent if you want. Here's how you can go about it.

Cut stems of your cacti and succulents are used to grow new plants.
https://www.pexels.com/photo/person-cutting-a-stem-6764325/

1. Ensure that the cut stem is healthy with fresh-looking leaves.
2. Keep the stem as is in your room with enough indirect sunlight for a couple of days or more. That way, it will callus, and the part where it was cut will become dry, keeping rot at bay.
3. Pour damp cactus or succulent soil into a fresh pot. Just to refresh your memory, cactus soil has more inorganic sediments, whereas succulent soil has more organic ones.

4. Insert the cut stem into the soil with the callused part in the soil. If there are leaves on the sides, it may not go in. Pluck them off, but only those near the base.
5. Place the pot in indirect sunlight, say on a porch with a roof.
6. Water the soil and wait for a couple of weeks. From here on, you only need to mist the soil if it dries up, which may happen after a week or so.
7. Continue misting until you can see roots forming in the callused stem. It may take a month or even a whole year.
8. Once roots have begun to form, you can water and care for your newly grown cacti and succulents, as shown in Chapters 4 and 5.

You can try changing the rooting medium from soil to water, but there is a much higher chance that the stem may rot while submerged in water. For a better success rate, stick with cactus and succulent soil.

Despite healthy soil containing the required nutrients and enough water, is your stem still showing no roots after several months? You can try to kickstart the process by using rooting hormones. They are available in powdered or liquid form, consisting of natural auxins that prompt your stem to initiate root growth.

Stem cuttings produce genetic clones of your cacti and succulents, but that's not the only way that you can propagate or clone your plants. Leaves can also be used for the same purpose.

Leaf Cuttings

Similar to stem cuttings, the cut leaves of your succulents are used to grow new plants. This type of propagation isn't applicable to cacti since they don't contain any leaves. Cloning through leaf cuttings is one of the fastest types of propagation since you can expect new roots in around two months. The process is just like that of stem cuttings, with barely any deviations.

1. Choose the succulent that you want to duplicate. It must contain enough moisture in the sense that its leaves don't feel dry and shriveled.
2. Pluck a fresh leaf from the bunch, one that feels healthy and fleshy. A simple pull won't do the trick. You will need to twist the base of the leaf slightly before pulling it out.
3. You may notice some wetness on the plucked part, which is good since it implies that it has enough stored water to be fresh and

healthy. Before you can push it into the soil, you will need to dry it. Let it lie on a dry surface for about two days, at least until you can notice calluses at the base. It may even take up to five days.

4. Pour the succulent soil into a pot and keep it damp before inserting the leaf, callus down, into it.
5. Place the pot in indirect sunlight and wait for a couple of weeks.
6. Take the leaf out and check the base. Can you see small threadlike roots down there? If not, you may need to wait another week.
7. After another few weeks, you will notice small versions of your succulent plant growing from those roots.
8. There will come a time when the original cut leaf falls off. That is when you have to unearth those baby succulents and plant them in a fresh drainage pot with a hole in the bottom.

If roots or baby succulents aren't developing, you may try dipping the callused part of the leaf in rooting hormones. Water rooting medium isn't recommended for leaf cuttings since they may quickly rot.

Division Propagation

Remember those sci-fi films of yore where a human is symmetrically cut into two parts, after which the symmetrical parts are regenerated to form two of the same human being? That is somewhat how division propagation works in cacti and succulents. You simply need to divide a single plant into multiple segments. Each segment contains all the necessary parts of the plant, right from its leaves down to its roots. Here are the basic steps of division propagation for your cacti and succulents.

1. Take a cactus or succulent and separate it from the soil. In many plants, you can easily pick out individual, divided segments with seemingly isolated parts. In those where the segments aren't readily apparent, you will need to tear the plant into small clumps.
2. Pour cactus or succulent soil into a drainage pot and dampen it a bit. Dig a long hole in the middle.
3. Insert the plant segment/clump into the hole and cover it with soil in such a way that the roots lie beneath the surface.

Eventually, a new cactus or succulent with the same genetic makeup will start growing in the pot.

A Point to Note: It is not possible to divide all species of cacti and succulents. Your plant needs to have one of three things to qualify for

division propagation.
- It should have multiple tiny stems instead of a single stump. It is easier to separate the roots in that way.
- Has your succulent given rise to smaller (baby) succulents on the side? They are called offsets, and they are perfect for division propagation.
- Have you planted two or more cacti in the same pot? Once fully grown, it is a great time to separate them into different pots through division propagation.

Handle the separate clumps with care, and make sure that none of their parts fall off during the transfer.

Now that you know the steps of every propagation technique and have mastered cloning cacti and succulents like an expert gardener, it's time to learn how to maximize your yield.

Maximizing Yield with Propagation

Do you want a good number of seeds or baby succulents from a single round of propagation? Do you want your yield to be lush and healthy, as if your cactus has been picked afresh straight from the desert? Follow these tips to maximize your yield during and after propagation.

- **Ensure Proper Nutrition**

Cacti and succulents get most of their nutrition from the soil itself, and since they thrive in a harsh, dry environment, you don't need to fertilize them as often as other plants. Fertilization only ensures that the plant can adapt to your home environment more easily, along with increasing the possibility of growing offsets if they are able to. Giving them the required nutrients once a year is enough.

For potted succulents, you may feed fish emulsion or manure tea. Ideally, any 5-10-5 fertilizer will work (for both cacti and succulents), where five is the amount of nitrogen, 10 is phosphorus, and the last five is the level of potassium. Of these, nitrogen ensures the healthy development of stems and leaves, whereas phosphorus focuses on the growth of flowers and roots. Make sure you dilute the fertilizer in some water because a concentrated dose may cause rot.

- **Provide Sufficient Exposure to Light**

Cacti and succulents can live for long periods without water and nutrition, but they need light to survive for even a short while. It's a no-brainer because these plants have adapted to drought areas with scorching heat and light. It is recommended that you keep your pot exposed to direct sunlight for at least six to seven hours each day. For outdoor planting, pick a spot where sunlight falls most of the day.

If you are planning to plant indoors in some dark corner of the room, then pot a plant that can survive without much light exposure. For example, the snake plant can flourish without direct sunlight in a semi-dark room for a long time. Its delightful appearance also improves your interior decor, and it purifies the air.

However, suppose you are bent on planting a light-requiring cactus or succulent indoors where there isn't sufficient sunlight. In that case, you may use artificial light for the purpose. You will need a mixture of fluorescent light and incandescent light in the ratio of 10:1. It means if you have a 10-watt fluorescent lamp, you will also need a 1-watt incandescent bulb. Keep these light sources on for 16 hours each day.

- **Keep the Temperature in Check**

You may think that cacti and succulents only survive in high-temperature environments. That is not true. Did you know they can do well in temperatures as low as 40°F? That is because the nights in the arid lands are often cold, so they have adapted to survive in varied temperature conditions. Problems may arise if the temperature falls below 30°F.

Do you have prolonged winters with lots of snow in your area? Don't be disheartened, as you can still grow and care for your plants in extreme cold. Place a burlap film covering the plant, ensuring it covers it completely. That will protect it from the direct chill while letting in enough direct sunlight through the pores.

Additionally, avoid watering your cacti and succulents in the winter because they exhaust their supply slower than usual. Pour water only if the soil has dried out or the plant has hardened due to lack of water.

- **Maintain the Required Humidity**

Cacti and succulents, like other plants, require humidity to survive, but they need it in smaller amounts. Most other plants have a survival range of 60% to 90% humidity, whereas succulents can thrive in even 10% humidity. Nevertheless, the recommended range for most cacti and succulents is around 40% to 60% humidity.

The amount really depends on one important factor – the thickness of the leaves. The thicker the leaves, the less humidity is needed. But how do the cacti with less thick leaves get humid in a desert, you may ask. They collect it from the fog. If your neighborhood doesn't see much fog during the year, you can give your plant what it needs through the misting process. Simply mist the leaves with filtered water.

- **Ensure Proper Spacing**

If you are a seasoned horticulturist, you may know how critical spacing is for efficient succulent growth. Are your garden cacti planted too close to each other? They will use up more water and nutrients than usual and may need more than six hours of sunlight.

The thing is, when plants are clumped too close together, they tend to compete for the basic requirements. And where there is competition, there's a winner and a loser – fundamentals of Charles Darwin's survival of the fittest. Thus, some of your plants in a crowd may thrive while others may wither away.

It is recommended that you space your cacti and succulents at least two feet apart. Large plants, like saguaro, which can grow quite wide and tall, will need more space of around four to five feet between them.

In short, to maximize the yield of cactus and succulent propagation, you must expose them to six hours of direct sunlight, keep the temperature above 40°F, and ensure two to three feet of spacing between two plants. Nutrition and humidity can be naturally acquired, but you may provide your plants with them to boost healthy growth.

Chapter 8: Pest Control, Disease Management, and Other Challenges

Just like humans need adequate care and disease management to stay healthy, succulents and cacti require pest control and proper disease management to remain healthy and thriving. Although cacti thrive well while there is limited water availability, they still need protection against pests and several diseases. These pests and diseases can affect their distinct appearance and compromise their growth.

Mealybugs, scale insects, whiteflies, spider mites, vine weevils, and caterpillars are commonly found pests that affect succulents and cacti. These pests primarily feed on plant tissue, disrupting the nutrient distribution within the plant and even transmitting harmful microorganisms that develop into diseases. Identifying and getting rid of pests early on is vital for your plants to survive.

Several fungi, viruses, and bacteria also affect cacti and succulents. These microorganisms cause diseases including leaf spot, stem rot, root rot, and other fungal infections. Most of the time, cacti's main culprits are poor conditions such as high humidity, poor air circulation, excessive fertilizer use, and overwatering.

Avid gardeners recommend using a holistic approach where prevention is the first line of defense. Providing the proper humidity, water, and sunlight and keeping the soil well-drained will ensure your cacti flourish and stay protected from diseases. Besides giving your plants a suitable growing environment, inspecting them regularly for any signs of disease or pest infestation is necessary.

A reasonably novel approach called Integrated Pest Management (IPM) is used chiefly to address pest infestations and diseases. This management method combines cultural, mechanical, biological, and chemical controls. Natural methods include pruning and handpicking pests. Chemical control involves using insecticides and fungicides, and mechanical methods involve using specific equipment to avoid harmful effects.

Common Pests That Affect Succulents

There are several pest species capable of disrupting the growth of cacti.
https://pixabay.com/photos/leafhopper-insect-macro-nature-562118/

There are several pest species capable of disrupting the growth of cacti. The detrimental effects a pest infestation might have on the plant depends on the type of pest and the severity of the infestation. Here are some common problems, how to identify them, and ways you can prevent this infestation.

Mealybugs

These small, soft-bodied insects are covered in a white, waxy, cotton-like substance that provides protection. They cluster on the plant's stems, base, and leaves. These bugs are oval and can range from a millimeter to four millimeters. With multiple legs, mealybugs have a white-colored waxy outer coating. This coating gives mealybugs their distinct cotton-like appearance. You'll see clusters of cotton-like substances on the leaves, stems, and base, where the plant sap is easily accessible.

Scale Insects

These tiny, oval-shaped insects can attach themselves to leaves and stems, sucking the plant sap and thriving unnoticed. They come in various colors, depending on the scale of insect species, but all have an outer soft body. Scale insect infestation in your cacti garden can turn them yellow or brown. These insects also feed on the plant's sap, sucking their life force (sap) and making your precious plants prone to other diseases. The cacti leaf can also appear as having a sticky appearance.

Aphids

These are small, soft-bodied insects resembling a pear in appearance. Aphids have long antennae and wings and come in different colors, including yellow, green, pink, and brown. These insects infest areas of new growth or the underside of leaves, making them difficult to identify. They also feed on the plant's sap, sucking the life force out of your prickly cacti. Like scale insects, aphids produce honeydew that attracts ants and other insects, increasing the chances of developing fungal infections.

Spider Mites

While spider mites seem like any other insect, they are arachnids, a subcategory of invertebrate animals. Spider mites are primarily brown, red, or yellow and are nearly impossible to identify without magnification tools. Like spiders, they have eight legs and feed directly on plant leaves. The damage spider mites cause results in yellowing of the area. Over time, the plant stops growing and declines even further.

Thrips

These tiny, slender insects have fringed wings that measure less than one millimeter. Thrips are yellow, black, or brown, have elongated bodies, and appear all over the plant. While feeding on the plant's tissue, thrips leave a silvery streak on the leaves. Slowly, your cacti will show distorted or discolored growth. Thrips can carry viral infections and spread disease to other plants.

Whiteflies

These tiny flying insects have white-colored wings containing a powder-like substance released during flight. Their body resembles a moth and has four wings. They reproduce under the leaves, hidden from plain sight. Similar to other insects, whiteflies feed on plant sap and cause wilting. They can also trigger sooty mold growth in the plant.

Fungus Gnats

Gnats are dark-colored flies that look similar to fruit flies. They are more active around the soil surface of potted plants. Their long, slender bodies make them agile flyers. Adult fungus gnats do not cause significant damage to the plants. However, their larvae feed on organic matter in the soil, including plant roots and tender root hairs, which can affect the plant's overall health and growth.

Root Mealybugs

Although root mealybugs look similar to regular mealybugs, this type thrives underground and feeds on the plant's roots. It's nearly impossible to detect their presence until their population becomes massive. Infected cacti show wilting and stunted growth.

Leaf Miners

Leaf miners are mostly larvae of insects like moths, flies, and beetles that feed on the plant's tissue, leaving holes in the leaves. They are worm-like creatures and create tunnels or trails on the leaf surface. These miners also vary in color, ranging from brown to white.

Vine Weevils

These beetles have a distinctive snout and don't fly like regular beetles. Weevils go unnoticed during the day as they are nocturnal.

Identification: Vine weevils are small, flightless beetles with a distinctive "snout." They are nocturnal and often go unnoticed during the day. Vine weevils prefer nutrition from the roots but can also feed on leaves, leaving semicircular notches.

Leafhoppers

Leafhoppers are insects with intricate patterns on their bodies and long hind legs that allow them to jump. They can also cause wilting, stippling, or yellowing leaves. Severe infestations can even inhibit new growth.

Regular inspection, maintaining hygiene, providing the best growth conditions, and controlling pests are the main elements that need to be taken care of. Knowing about diseases and pests is crucial to prevent them from happening. Lastly, start with the least harmful method to protect the cacti and succulents when identifying the pest. For example, if there are a few visible pests, use natural predators like beetles and harvestmen to keep the pest population at bay without affecting the plant. Lastly, when using insecticides, always use controlled amounts and follow instructions carefully for maximum results.

Integrated Pest Management Techniques

This comprehensive pest management technique is one of the most sustainable approaches toward pest control in cacti and succulents. The method aims to reduce insecticide use and incorporate several pest control approaches for an impactful output. Here are the strategies used collectively in integrated pest management.

This comprehensive pest management technique is one of the most sustainable approaches toward pest control in cacti and succulents.

https://pixabay.com/photos/slug-policeman-lego-garden-pest-1535140/

Cultural Control

This practice is a foundational pillar of pest management, focusing on fostering an environment where plants can thrive and build resistance against pests and diseases.

When purchasing the cacti and succulents, inspect them thoroughly for any signs of disease or pests to avoid bringing home infested plants.

Water the plants adequately and avoid overwatering to prevent the development of root-related diseases.

Before planting, ensure the soil is well-draining and the soil mix has adequate fertilizers to keep the cacti nourished. You can let the soil dry out between watering to discourage pest infestations.

Keep your beautiful cacti in places with ample sunlight and ventilation. They won't require direct sunlight, so one of the best places is near windows. Ventilation is crucial to prevent humidity build-up. Humid environments promote the development of fungal infections in plants and provide a feasible environment for pest infestation.

When adding new cacti to your collection, keep the newly bought plant isolated for a few weeks, ensuring no pest or disease infestation could affect other plants.

Mechanical Control

It involves the physical removal of pests from plants. The following mechanical control methods are used to get rid of unwanted pests.

Handpick pests on your cacti and succulents, especially on the undersides of leaves and along the stems, if any. In cases of pest infestation, use a soft brush or a cotton swab dipped in rubbing alcohol to wipe off pests like mealybugs and aphids.

Most pests infest specific areas, like the undersides of the leaves, and should be permanently removed by pruning. Use sharp and sterilized pruning shears to prevent further spread.

Biological Control

The biological method in integrated pest management promotes the use of natural techniques to keep the pest population at bay. Ladybugs and certain beetles can be used to prevent the infestation of mites and several other pests. Similarly, adding nematodes to the soil can control pests like mealybugs and fungus gnats.

Chemical Control

Although chemical control methods are included in IPM, using this as a last resort is strongly advised after you've tried other pest control methods.

Start with the least toxic pesticide and avoid broad-spectrum insecticides that can harm insects living in symbiosis with the cacti. Pay attention to the safety precautions, timing, and application rates, and always use protective gear before application. Instead of spraying pesticides all over the plant, keep them confined to specific areas of infestation, reducing their adverse effects on the cacti.

Adopting the controls mentioned above will keep plants healthy and make pest management effective. Besides using these techniques, make it a habit to inspect your plants regularly so you can act early and stop pest infestations in your cacti collection.

Understanding Succulent Diseases

Although cacti are resilient plants and can thrive in harsh environments, they are still prone to various diseases. Common cacti diseases can result from fungal, bacterial, and viral infections. Knowing the disease's signs and symptoms is necessary for early intervention and effective management.

Fungal Infections

Fungal infections occur in humid environments or when the plant has been overwatered. Some commonly occurring diseases in cacti include the following.

Root Rot

This fungal infection causes wilting, changing the leaf color to brown or yellow. Slowly, the plant loses its ability to grow, and the root system starts turning brown. The root system also becomes rotten and produces a distinct and musty smell. The first sign is the change in the root color, which turns from white to dark.

Stem Rot

The stems affected by stem rot turn black or brown, become sunken, and can even ooze sap. The area gets mushy, which, on closer examination, can reveal discolored tissues. Fungal spores are also visible on the stems when the disease is fully developed.

Leaf Spot

The leaves in this fungal infection become irregularly shaped and develop distinct margins. Concentric leaf patterns can be seen on the leaves, and they turn brown or gray. Fungal spores can also be identified on the leaves when the fungal infection has fully developed.

Bacterial Diseases

Bacterial Soft Rot

Soft lesions start developing on the stems, which can quickly become slimy and mushy. Bacterial soft rot, when developed, has a foul odor that becomes intense as the infection turns severe. Dissecting the affected area can reveal disintegrated tissue that has already lost its functionality.

Bacterial Crown Rot

Bacterial infection limited to the area where the stem meets the roots is known as bacterial crown rot and can damage the plant's integrity. The stem becomes weak, and eventually, the plant collapses. Yellowing or wilting becomes evident as the stem is unable to transport nutrients from the stem throughout the plant.

Viral Infections

Like any other viral infection, they don't have a specific cure. Once a plant is infected, the virus remains in it for its lifetime. Common viral infections include the following:

Cactus Virus X

Mosaic patterns or dark green rings can be seen in cacti affected by cactus virus x. The plant will show stunted growth when infected, and the flowers will fail to open or become deformed.

Acting promptly is crucial during the diagnosis and treatment of a disease. Isolating the infested plants, preventing further spread, providing the right growing environment, and avoiding overwatering are some practical techniques that are beneficial in controlling the disease symptoms.

For most fungal and bacterial infections, isolating and treating the plant with a feasible medication can prevent further spread to nearby plants. However, destroying the infected plant might be necessary for viral infections if the viral infection is severe.

Management of Succulent Infections

Fungal Infections

Improve Drainage: Plant cacti and succulents in well-drained soil. Stagnant water within the soil can create good conditions for fungal growth.

Ensure your cacti and succulents are planted in well-draining soil. Avoid soggy conditions that promote fungal growth and root rot. Use a potting mix specifically designed for cacti and succulents.

Reduce Humidity: Maintain good air circulation around your plants and avoid high humidity. Fungal pathogens thrive in humid environments, so ensure proper ventilation in indoor spaces and avoid overcrowding plants.

Remove Infected Plant Parts: As soon as you identify the first sign of a fungal infection, immediately prune the affected area, limiting the spread of the disease to unaffected parts. When removing the diseased area, use clean pruning shears and disinfect between cuts.

Apply Copper-Based Fungicide: Severe fungal infections require the application of a copper-based fungicide. This fungus-killing substance is effective against a variety of fungal pathogens. Always apply the fungicide on a small area to check for adverse effects, and then use the fungicide on the main affected area according to the instructions.

Bacterial Infections

Practice Sanitation: Keep your growing area clean by removing fallen leaves and debris, as they can harbor bacterial pathogens. Regularly clean your gardening tools with rubbing alcohol or a bleach solution to prevent the transmission of bacteria between plants.

Isolate Infected Plants: Immediately isolate plants that show signs of bacterial infection to prevent the spread to healthy ones. Monitor the isolated plant/s closely and avoid handling other plants before washing your hands and disinfecting tools.

Improve Air Circulation: Maintain adequate spacing between plants to enhance air circulation. Air circulation reduces humidity and decreases the chances of developing bacterial infections.

Remove Infected Plant Parts: At the first sign of bacterial infection, remove and destroy the affected parts, including leaves, stems, or the entire plant if necessary — disinfect pruning tools between cuts.

Viral Infections

Prevention Is Key: Virtually every viral infection affecting plants has no definitive cure. There is no cure for viral infections in plants. The best approach is prevention. Purchase plants from reputable sources and inspect them thoroughly for any signs of viral diseases before adding them to your collection.

Quarantine New Additions: Isolate new plants for a few weeks to observe any symptoms of viral infections before integrating them with your existing collection.

Sanitize Tools and Hands: Disinfect gardening tools by handling different plants to prevent the transmission of viruses. Wash your hands thoroughly before and after handling plants, especially if you suspect an infection.

Remove and Destroy Infected Plants: If a plant shows clear signs of viral infection, such as mosaic patterns or rings on the leaves, remove and destroy the infected plant to prevent the spread to other healthy plants.

In all cases, acting fast when you notice any signs of disease is crucial. Early detection and quick action can significantly improve the chances of successful disease management. For severe or persistent infections, consider seeking advice from a plant expert or horticulturist who may provide additional guidance for your situation.

Remember that maintaining a healthy and disease-free environment for your cacti and succulents is the most effective approach to preventing diseases. Regularly monitor your plants, provide proper care, and maintain good sanitation practices to ensure their well-being.

Management of Common Challenges

Environmental Challenges

Temperature Extremes: Cacti and succulents are adapted to various temperature ranges, but extreme heat or cold can stress the plants. Frost and freezing temperatures, particularly, can damage some species, leading to tissue damage and discoloration.

Inadequate Lighting: Insufficient sunlight can result in elongated and weak growth. On the other hand, too much direct sunlight can cause sunburn and damage to the plant's tissues.

Humidity Fluctuations: Fluctuations in humidity levels can stress cacti and succulents, especially those that prefer dry conditions. High humidity can increase the risk of fungal diseases, while low humidity can lead to dehydration and wilting.

Poor Air Circulation: Inadequate air circulation can increase plants' humidity and make them more susceptible to diseases like powdery mildew and fungal infections.

Overwatering and Underwatering

Overwatering: One of the most common mistakes in succulent care is overwatering. Cacti and succulents are adapted to survive in arid conditions and prefer infrequent, deep watering. Overwatering can lead to root rot and other fungal diseases.

Underwatering: Underwatering can cause cacti and succulents to become dehydrated and may result in wilting, shriveling, or the appearance of dry, crispy leaves.

Nutritional Deficiencies

Nitrogen, Phosphorus, or Potassium Deficiencies: Like all plants, cacti and succulents require essential nutrients for healthy growth. A lack of nitrogen, phosphorus, or potassium can lead to stunted growth, yellowing leaves, and poor overall health.

Plant Stress

Physical Damage: Accidental damage to the plant's stem, roots, or leaves can create openings for pathogens to enter and cause infections. Avoid rough handling or contact with sharp objects.

Transplant Shock: Transplanting cacti and succulents can cause stress, particularly if the roots are disturbed. Transplant shock may result in wilting, discoloration, or slower growth as the plant adjusts to its new environment.

Environmental Stressors: Exposure to extreme weather conditions, such as strong winds, intense sunlight, or sudden temperature changes, can stress cacti and succulents, affecting their overall health and appearance.

To effectively address these challenges:

- Know your plant's specific requirements and adapt care practices accordingly.
- Provide the right light, water, and humidity for the specific species.
- Monitor your plants regularly for signs of stress, pests, and diseases.
- Take preventive measures to protect your plants from extreme environmental conditions.
- Address any issues promptly to prevent further damage or the spread of problems.

By understanding and addressing these challenges, you can create a suitable and supportive environment for your cacti and succulents, promoting their optimal health and ensuring their longevity in your care.

Chapter 9: Companion Planting with Cacti and Succulents

Companion planting is an excellent gardening concept where cacti and succulents are grown with other compatible plant species to maximize the garden's health. Just like the concept of symbiosis, companion planting is based on the idea that certain plant species, when grown together, will enhance growth, prevent pests, and promote nutrient uptake. This technique has been used since ancient times and is categorized as one of the most sustainable approaches to gardening.

A harmonious garden can be created by planting companion plants that support each other's growth and well-being. Some plants can produce pest-mitigating chemicals, whereas others can benefit from nitrogen fixation in the soil.

Companion planting is an excellent gardening concept where cacti and succulents are grown with other compatible plant species to maximize the garden's health.
https://www.pexels.com/photo/potted-flowers-placed-in-cozy-room-4813268/

Although companion planting has numerous benefits, it's necessary to understand that each plant has specific requirements. Not every plant combination will work or be mutually beneficial. Therefore, studying plant species and doing your research before implementing companion planting techniques is crucial. Still, companion planting is an excellent way for gardeners to enhance biodiversity, reduce dependence on pesticides and fertilizers, and make the plants resilient. When done right, this symbiotic companionship of plants can go a long way toward reducing costs for pesticides and other problems in your garden.

The Purpose of Companion Planting

Although cacti and succulents can withstand harsh conditions and don't need intensive maintenance, companion planting can still serve several purposes. Companion planting fosters a sustainable and holistic approach to gardening as it can naturally prevent the spread of pests and diseases.

Pest Control

Aromatic herbs like rosemary, lavender, and thyme can be planted alongside cacti, deterring harmful insects and pests. Planting these companion plants creates a pest-free ecosystem that reduces the need for chemical pesticides.

Biodiversity

Planting compatible species with your cacti and succulents promotes biodiversity. Several companion plants can also attract beneficial insects, pollinators, and other microorganisms that can nurture the garden's overall health. For example, flowering companion plants can enhance pollination rates in other plants by attracting butterflies.

Soil Health

Companion plants like legumes and beans encourage the development of nitrogen-fixing bacteria, which enrich the soil with nitrogen. Cacti and succulents benefit from the nitrogen reserves in the soil, decreasing the dependency on chemical fertilizers. Improved soil fertility will eventually promote healthier growth.

Visual Appeal

Adding visual interest and diversity to your cacti is also an added benefit of companion gardening. You can choose from a wide range of plants, herbs, and other ornamental plant species to make an aesthetically pleasing and harmonious display.

Microclimate Modification

Although cacti can tolerate drought-like conditions and high temperatures, the intense summer heat can affect their health. Planting companion plants that can provide shade to the cacti during the intense summer heat creates better growing conditions and lets you create microclimate conditions.

Natural Pest Predators

Several companion plants attract ladybugs, plant-friendly beetles, and lace wigs, which feed on common pests. These natural pest predators can minimize the dependency on insecticides or chemical interventions.

Improved Pollination

If you have a full-size garden, choosing plants like the African daisy, Mexican daisy, fortnight lily, red valerian, etc., with your cacti will increase pollination, keeping the garden thriving. Marigolds and sunflowers can also be used in specific areas to promote pollination.

Weed Suppression

Companion plants like mint, artemisia, catmint, and stonecrop prevent weeds. These plants don't leave enough space for the root systems of weeds to take root and thrive. Dense plantings can physically shade the soil in extremely hot environments, further preventing weeds from taking hold.

Space Optimization

You can easily elevate the garden's aesthetics and use every nook and cranny by planting visually striking plants. For example, growing a tall or climbing plant with cacti is an excellent choice if you have limited space. The tall, rising plants will provide your cacti enough shade, protecting them from direct sunlight. Similarly, virtually any available space can be used accordingly, creating a biodiverse ecosystem.

Having tons of benefits, companion planting can pave the way toward sustainable gardening, keeping the plants healthy and robust. Imagine creating a garden of cacti and succulents, surrounded by magnificent flowering companion plants that can attract plant-friendly insects, herbs, and aromatic plants to ward off pests and plants like alfalfa and clovers, keeping the soil rich in nitrogen. All these benefits are possible through companion planting. However, it will need research and knowledge about companion plants compatible with cacti and succulents.

Suitable Companion Plants

Selecting suitable plants involves considering their compatibility regarding growth habits, nutrient needs, pest management, and mutual benefits. Here are some examples of compatible companion plants and the reasons for pairing them:

Selecting suitable plants involves considering their compatibility regarding growth habits, nutrient needs, pest management, and mutual benefits.
https://pixabay.com/photos/chilli-lavender-red-mauve-orange-265756/

Succulents

Agaves, sedums, and echeverias are some succulent companion plants for cacti. Succulents and cacti have virtually similar growth requirements, making them ideal plants. Their sunlight, soil, and nutrient requirements are the same. When planted together, Cacti and succulents create a visually striking and cohesive desert landscape. Veteran cacti gardeners recommend planting succulents near cacti, adding color, texture, and depth to the landscape.

Creeping Plants

For lush foliage, consider low-growing and spreading plants like creeping thyme, ice plants, and various sedum species to add to your garden. These plants can provide excellent soil cover around the cacti, serving as a natural mulch that retains moisture, regulates temperature, and prevents weed growth. They also create a living carpet that adds visual interest and complements the upright forms of cacti. These creeping plants can also provide a feasible habitat for plant-friendly insects and pests, limiting harmful pest infestation.

Wildflowers

Desert-adapted wildflowers like desert marigolds (Baileya multiradiata), desert bluebells (Phacelia campanularia), and globe mallow (Sphaeralcea spp.) can bring bursts of vibrant color to your cactus garden. These flowers attract pollinators, such as bees and butterflies, which help improve pollination for cacti and wildflowers.

Aloe Vera

Aloe vera is not only a popular companion plant for cacti due to its similar water requirements but also because of its practical uses. Aloe vera's succulent leaves contain a soothing gel that can be used for various skin ailments, making it a functional addition to your garden. Planting aloe vera near cacti provides some shade and protection to the cacti while offering a source of natural skincare.

Lavender

Lavender has such a pleasant fragrance and can also attract beneficial insects. Planting lavender near cacti can create a sensory experience in your garden and support pollinators like bees. The tall, slender stems and purple blooms of lavender contrast beautifully with the spiky, textured appearance of cacti.

Yucca

Yucca plants share a similar arid habitat with cacti and can make excellent companions in a desert-themed garden. Their upright growth habit and dramatic flower spikes add visual interest and height to the landscape. Yuccas can create a striking backdrop for cacti while contributing to the overall desert aesthetic.

Herbs

Certain herbs that thrive in dry conditions, such as rosemary, sage, and oregano, can be planted alongside cacti. These herbs add aromatic scents to your garden and also serve culinary and medicinal purposes. Their drought-tolerant nature and compatibility with cacti make them versatile companions.

Rocks and Gravel

Incorporating decorative rocks and gravel into your cactus garden can mimic the natural desert environment and also have practical benefits. Rocks and gravel help regulate soil temperature, prevent erosion, and enhance drainage. They also create a visually appealing hardscape that complements the unique textures of cacti.

When planning your companion planting arrangement, consider the specific needs of each plant, including water, sunlight, soil, and space requirements. Group plants with similar needs together to ensure they thrive and harmonize. Regular monitoring, proper spacing, and adjusting care routines as needed will contribute to a successful and visually captivating cactus garden with well-chosen companion plants.

Checking Plant Compatibility

Considering compatibility in terms of light requirements, water needs, growth habits, and soil preferences is essential to create a successful and harmonious garden environment. Each plant species has unique characteristics and preferences, and understanding these factors helps gardeners make informed decisions about plant placement and combinations. Here's why compatibility matters:

Light Requirements

Different plants have varying light preferences, ranging from full sun to partial shade or full shade. Planting species with similar light requirements together ensures that all plants get the appropriate amount of sunlight. Placing shade-loving plants in direct sunlight or sun-loving plants in the shade can lead to stress, wilting, or poor growth. By grouping plants with similar light needs, you optimize their overall health and vitality.

Water Needs

Water requirements can vary widely among plant species. Some plants are drought-tolerant and prefer infrequent watering, while others need consistently moist soil. Pairing plants with similar water needs ensures that all plants receive the appropriate amount of water, preventing overwatering or underwatering. Overwatering leads to root rot and other water-related issues, while underwatering can cause plants to wilt and weaken.

Growth Habits

Different plants have different growth patterns, such as height, spread, and growth speed. Managing the growth habitat is essential to prevent overcrowded plants and make the existing plants fight for resources. For example, if you pair a fast-growing plant with cacti or succulents, the aggressive-growing plant can easily outcompete nearby plants, using up all the soil resources themselves. Selecting plants with complementary growth habits ensures that all plants have sufficient space to develop fully without restricting one another's growth.

Soil Preferences

Each plant has varying soil preferences, which include different pH levels and nutrient requirements. Soil humidity and soil texture. Some plants prefer well-draining soil, whereas others prefer thriving in moist and loamy soil. Placing plants with similar soil preferences together helps create an ideal growing environment for each species. Additionally, certain plants can improve soil conditions for others through nitrogen fixation or soil aeration.

Plant Compatibility

Choose plants with similar growing requirements regarding sunlight, water, and soil preferences. Avoid pairing plants with wildly different needs, which can lead to resource competition and hinder their growth. For example, pair sun-loving or shade-tolerant plants together to ensure they flourish in their respective environments.

Growth Habits

Another issue to think about is the growth habits of the plants and how they will interact physically in the garden. Tall plants can provide shade or support for shorter plants. For instance, tall sunflowers can offer shade to cacti succulents. Lower-growing lettuces or provide a sturdy trellis for climbing beans.

Nutrient Needs

Pair plants with different nutrient needs together to ensure efficient use of soil resources. Some plants, like legumes, fix nitrogen in the soil, benefiting nearby plants that require nitrogen for healthy growth. For example, grow nitrogen-fixing beans or peas alongside growing succulents and cacti to provide them with the required nitrogen to thrive.

Pest-Repelling Properties

Use companion plants that deter pests or attract beneficial insects. Certain herbs, flowers, or aromatic plants have natural pest-repelling properties, which can help protect nearby susceptible crops. For example, planting marigolds can repel pests and insects, preventing infestations that could potentially affect the cacti.

Attracting Pollinators

Include plants that attract pollinators like bees, butterflies, and other beneficial insects. Flowers with abundant nectar and pollen can increase pollination rates and fruiting. For instance, planting lavender or borage near cacti will increase their pollination rate.

Succession Planting

Pair plants with different growing seasons or maturity rates to keep the surroundings in control. For example, you can plant fast-growing crops like radishes or lettuce alongside succulents when you need to provide shade, limit soil erosion, and keep weed production at bay. These faster crops are harvested, allowing more space and resources for the slower ones to expand.

Aesthetic Appeal

Consider the visual impact of companion plant combinations. Combine plants with contrasting colors, textures, and heights to create visually appealing and harmonious garden displays. For example, pairing purple basil and bright yellow marigolds with cacti and succulents can create a striking and vibrant color contrast.

Native and Regional Plant Choices

Select native or regionally adapted plants for companion planting to support local ecosystems and encourage beneficial wildlife, such as native pollinators. Native plants are often well-suited to the local climate and require less maintenance.

In summary, pairing suitable companion plants with your cacti and succulents involves thoughtful planning, taking into account plant compatibility, nutrient needs, growth habits, pest-repelling properties, and aesthetics. By creating a diverse and balanced garden ecosystem, gardeners can maximize productivity, reduce pest pressures, support pollinators, and enhance the overall health and beauty of their gardens.

By considering these compatibility factors, gardeners reap several benefits:

Efficient Resource Allocation: Pairing plants with similar requirements will minimize the waste of resources. Water and fertilizer can be applied more precisely, ensuring that each plant gets what it needs without excess or deficiency.

Reduced Maintenance Efforts: When plants are compatible, they are more likely to thrive in their environment, leading to fewer pest problems and less plant stress. This can reduce the need for intervention and maintenance efforts.

Aesthetically Pleasing Arrangements: Grouping plants with complementary growth habits, colors, and textures can create visually appealing landscapes or garden beds.

Enhanced Biodiversity: By selecting a diverse range of compatible plants, gardeners promote biodiversity and attract a wider array of beneficial insects and wildlife to the garden.

Overall, considering compatibility in terms of light, water, growth habits, and soil preferences is crucial for successful gardening. It enables gardeners to create balanced, sustainable, and visually pleasing garden environments where each plant thrives and contributes to the overall health and productivity of the space.

Companion Plant Recommendations

Pest-Repelling Companion Plants

Pest-repelling companion plants emit natural compounds or scents that deter common garden pests. Including these plants in your garden can help reduce pest populations and minimize the need for chemical pesticides. Some examples of pest-repelling companion plants are:

Marigolds: Marigolds are well-known for repelling aphids, whiteflies, and nematodes. Their distinct scent acts as a natural deterrent, making them effective companions for vegetables like tomatoes, peppers, and potatoes.

Nasturtiums: Nasturtiums are another excellent choice for pest control. They deter aphids, whiteflies, squash bugs, and cucumber beetles. Planting nasturtiums near cucumbers, squash, and tomatoes can help protect these vulnerable crops from common garden pests.

Nitrogen-Fixing Plants and Soil Enhancement

Nitrogen-fixing plants improve soil fertility by enriching it with nitrogen, an essential nutrient for plant growth. These plants have a symbiotic relationship with nitrogen-fixing bacteria, which convert atmospheric nitrogen into a form that is readily available to plants. Examples of nitrogen-fixing plants include:

Legumes (Beans, Peas, Lentils, etc.) are well-known nitrogen-fixers. They form nodules on their roots, where nitrogen-fixing bacteria reside. As these plants grow, they take nitrogen from the air and store it in the soil, benefiting neighboring plants like leafy greens, brassicas, and corn.

Pollinator Attractants

Pollinator-attracting companion plants play a vital role in garden ecosystems by attracting bees, butterflies, and other beneficial insects. These pollinators are essential for fruit set in many flowering plants. Examples of pollinator-attracting companion plants include:

Bee-Friendly Flowers: Flowers like zinnias, sunflowers, cosmos, and lavender are attractive to bees and other pollinators. Planting these flowers near fruiting plants like cucumbers, tomatoes, strawberries, and melons can enhance pollination rates and increase the yield of these crops.

Aesthetic Combinations

Aesthetic combinations focus on creating visually appealing garden displays by combining plants with contrasting colors, textures, and forms. Some examples of aesthetic combinations are:

Color and Texture Contrasts: Pairing plants with contrasting colors and textures can create striking visual displays. For instance, you can plant purple sage with yellow marigolds for a complementary color combination or red roses with white daisies for a classic contrast.

Container Gardening and Indoor Companions

Companion planting is not limited to traditional outdoor gardens. It can also be used in container gardening and indoor plant arrangements. Some examples of container and indoor companionship are:

Herbs and Vegetables: Planting herbs like basil, mint, or parsley alongside compact vegetables in containers can save space and provide fresh herbs for culinary use. For example, pairing cherry tomatoes with basil or salad greens with chives can be a practical and aesthetically pleasing combination.

Succulents and Cacti: Mixing different varieties of succulents and cacti in containers can create beautiful, low-maintenance displays for indoor and outdoor spaces. Combining these resilient plants' different shapes, colors, and sizes can result in eye-catching arrangements.

Care and Maintenance Considerations

While planning companion plantings, it's essential to consider each plant's specific care and maintenance needs. Some key considerations include:

Watering: Group plants with similar water needs together to ensure efficient watering and prevent over or under-watering. For example, pairing drought-tolerant plants like succulents with other low-water-need plants can simplify watering routines.

Pruning: Plan for growth habits and space requirements to avoid overcrowding and provide enough room for plants to grow without shading each other excessively. Regular pruning and trimming of plants help maintain the desired balance in the garden.

Pest Management: Consider the pest-repelling properties of companion plants when planning your garden layout. Including pest-repelling plants strategically throughout the garden can help naturally deter common pests, reducing the need for chemical pesticides.

By considering these factors and carefully selecting companion plants based on their benefits and compatibility, gardeners can create thriving and sustainable garden environments. Companion planting fosters biodiversity, enhances natural pest control, and adds aesthetic appeal to the garden, ultimately contributing to a more balanced and harmonious ecosystem.

Appendix: A-Z of Cacti and Succulents: Species Identification Reference

With thousands of species of cacti around the globe, there is at least one succulent for every letter of the alphabet. This appendix is an A-Z guide to some of the most interesting and easily identifiable cacti and succulents you might come across.

Acanthocalycium Glaucum

Scientific Name: Acanthocalycium Glaucum

Common Name: Acanthocalycium Glaucum

This takes a circular shape and often elongates into a single cactus. The stem grows up to 5.9 inches tall and around 3.15 inches in diameter. The Acanthocalycium Glaucum has 8 to 14 ribs, which are all covered by pruina for protection. The plant's spines are around 0.9 long and are clear at the base and dark at the ends. The Acanthocalycium Glaucum also has no central spine. It grows either red or yellow flowers that grow up to 1o inches tall.

Acanthocalycium Thionanthum

Scientific Name: Acanthocalycium Thionanthum

Common Name: Acanthocalycium Thionanthum

This cactus has a single habit that may branch out slowly as it grows older. It grows up to 4.7 inches long and 3.9 inches in diameter. The plant's stem is globular and is light green to bluish green and has 9 to 15 ribs. Younger cacti have black spines that turn yellowish as they age. The spines are long and sharp and are found in groups of 5 to 10 in each areola, along with one to four central spines. The Acanthocalycium Thionanthum grows bell-shaped bright yellow flowers and rarely grows orange or white ones.

Ball Cactus

Scientific Name: Parodia Magnifica

Common Name: Ball Cactus

This cactus has a ball-shaped to cylinder-like stem that grows up to 12 inches tall and 18 inches wide. The cacti of this variety can grow in clusters and frequently produce yellow, pink, red, or orange flowers. Younger cacti have white spikes that become yellowish-brown as they age.

Cabega

Scientific Name: Austrocephalocereus dybowskii

Common Name: Cabega

Cabegas are 3.9 inches in diameter and have cylinder-like stems. They grow in groups and are covered in white fluff. Each stem has around 20 ribs and two to three central spines, along with numerous tiny radial spines. Cabega flowers are white and bell-shaped.

Dutchman's Pipe Cactus

Scientific Name: Epiphyllum oxypetalum

Common Name: Dutchman's Pipe Cactus

This cactus can grow over 9.8 feet long and has several branches. The stems can grow anywhere between 6.5 to a whopping 19.6 feet long, causing them to flatten and lie sideways. They grow ample large white flowers and rarely produce fruit.

Echinopsis

Scientific Name: Echinopsis Thelegona

Common Name: Echinopsis

This cactus grows up to 7.2 feet tall and 3.9 inches in diameter. It produces a few shoots, and the stem has 12 to 13 ribs. Echnoposis is characterized by its areoles, which grow in hexagon-like patterns. The cactus has five to eight radial spines and one middle spine. In rare cases, Echiposis can exist with two middle spines. The plant grows white flowers that are around 8.6 inches long and 6.6 inches in diameter.

Fairy Castle

Scientific Name: Acanthocereus Tetragonus

Common Name: Fairy Castle

The Acanthocereus Tetragonus is among the easiest cacti to spot because it resembles a castle that a fairy would live in. The Fairy Castle has several stems, each of a different height. The average height of the stems is 3.2 feet. The stems have white spines along their ribs and grow at least a hundred branches and branchlets.

Giant Chin Cactus

Scientific Name: Gymnocalycium Saglionis

Common Name: Giant Chin Cactus

The Giant Chin Cactus is a single, huge stem with long spines. The stem looks like a somewhat flat sphere and is 7.4 to 15.7 inches in diameter. Most of the Giant Chin Cacti are 5.9 to 11.8 inches tall. However, they can grow up to 2.9 feet tall. They have one to three straight central spines and 10 to 15 curved radial ones. They produce red, globular fruits and funnelform blooms.

Hooker's Orchid Cactus

Scientific Name: Epiphyllum hookeri

Common Name: Hooker's Orchid Cactus

You could mistake the Epiphyllum Hookeri for a non-succulent at first glance because of its long foliage. However, if you look closely enough, you'll see the stems and spines of the cactus. The plant produces beautiful white flowers and grows up to 9 inches tall.

Johnson's Beehive Cactus

Scientific Name: Echinomastus Johnsonii

Common Name: Johnson's Beehive Cactus

This cactus has cylinder-like stems that grow up to 10 inches tall and 4 inches in diameter. It has countless bent spines that come in various colors, like purple, red, yellow, pink, and gray. Johnson's Beehive Cactus produces pink and yellow flowers that grow a little over 3 inches tall.

Koko

Scientific Name: Echinopsis Formosa

Common Name: Koko

Koko cacti are either globular stems with golden yellow spines that grow in clusters or cylindrical-shaped stems with creamy white spines that grow alone. Echinopsis Formosa grows up to 28 inches tall and 16 inches in diameter. They aren't quick to grow, so it will take them at least three decades to reach their full growth potential. Koko cacti grow yellow flowers.

Lady Finger Cactus

Scientific Name: Mammillaria Elongata

Common Name: Lady Finger Cactus

These cactus stems take the shape of fingers, with each growing up to 8 inches tall and 1.2 inches wide. Their spines are golden yellow, contrasting perfectly against the dark green stems. The cactus blooms with light pink, yellow, or white flowers.

Mammillaria Dixanthocentron

Scientific Name: Mammillaria Dixanthocentron

Common Name: Mammillaria Dixanthocentron

This cylindrical cactus has two to four yellow and soft reddish central spines and 19 to 20 white radial ones. It grows up to 11.8 inches tall and is 2.7 to 3.14 inches in diameter. It produces small, light red or pink flowers that are around 3.9 inches long. It also produces yellow or orange fruit.

Old Lady Cactus

Scientific Name: Mammillaria Hahniana

Common Name: Old Lady Cactus

This cactus is a natural work of art. It has a globular stem and often grows in clusters. The average single stem is 4 inches tall and 5 inches in diameter but can reach 10 inches in height over time. While the spikey fuzz ball looks interesting on its own, the violet-red flowers that blossom during summer create an astonishing crown-like effect.

Rattail Cactus

Scientific Name: Aporocactus Flagelliformis

Common Name: Rattail Cactus

This is yet another succulent that you can easily identify. It is characterized by its long, thin, snake-like stems, which grow up to 4 feet tall. The Rattail Cactus produces captivating purplish-red flowers but may sometimes bloom orange and pink ones.

Saguaro

Scientific Name: Carnegiea gigantea

Common Name: Saguaro

This giant succulent is what comes to most people's minds when they hear the word "cactus." It is how this plant is depicted in most cartoons and movies that are set in the Mexican or Arizona deserts. The cactus grows up to a staggering 50 feet in length and 30 inches in diameter. Some cacti grow one to five arm-like branches, while others grow none at all.

Thimble Cactus

Scientific Name: Mammillaria Gracilis

Common Name: Thimble Cactus

Many people prefer to keep Thimble Cacti as home plants due to their unique and decorative appearance and small size. They grow in clusters, with each growing up to 3 inches tall and 2 inches wide. Thimble Cacti produce white or pink flowers as well.

Zebra Plant

Scientific Name: Haworthia Fasciata

Common Name: Zebra Plant

This eye-catching succulent is easy to identify from miles away. It has glossy, long, and pointy dark green leaves, with creamy white stripes highlighting the veins, resembling zebra skin. The succulent produces yellow or white flowers at the top of the stem.

Conclusion

It's time to refresh your memory with a brief summary. In the biological sense, a cactus is a succulent, but gardeners and horticulturists define them as two different plants. Succulents have been evolving for millions of years, but they became a trend among humans soon after the birth of social media.

There are over 10,000 species of succulents, of which 1700 are cacti. So, it may be difficult for you to choose just one for your collection. Simply remember four important points: growing conditions, plant preferences, available resources, and the ultimate goal for the plant.

After selecting the right species, be very careful with the planting process. Plant two or more succulents as far apart as possible and prepare the planting area with suitable soil and allowance for drainage. If you follow all the best techniques, then there is a high chance that your plant will grow into the perfect specimen your heart desires. Also, since cacti and succulents are mostly ornamental, don't forget to arrange them in a visually appealing manner.

Then comes the most important part of nurturing these plants: watering, or rather, the lack of it. You need to take great care that you don't overwater any species of cacti or succulents. Use the right watering tools and techniques, take the temperature of the surroundings into account, and make sure to assess the moisture level in the soil before pouring in more water.

That said, it is absurdly simple to care for and maintain your cacti and succulents since they rarely ever need anything. You just need to remember a few significant points, like fertilizing options, seasonal care adjustments, pest and disease control, and essential care practices.

Furthermore, since cacti and succulents are primarily ornamental, you can prune them into appealing shapes and figures. While it's not as easy as pruning a bonsai tree since its leaves and stems are considerably thicker, it's not an exhausting process either. Pruning and cutting these plants has a side benefit, too. You can use the chopped parts to grow more plants of the same kind with a technique called propagation or cloning.

Multiplying and expanding your garden is an intriguing prospect, but you need to make sure that you have planted all the necessary companion plants. They not only beautify your backyard but also help in the growth of other plants.

It is recommended that you keep this book open at all times throughout the planting process to refer back to specific instructions, at least during your first few tries. Once you have retained all the information shown here and you have become used to the entire procedure, you can let go of the crutches and soar high and free into the realm of cacti and succulents, all on your own.

Here's another book by Dion Rosser that you might like

References

"11 DIY Homemade Plant Fertilizers (with Recipes)." *Gardening*, 21 May 2022, https://gardening.org/homemade-plant-fertilizer-recipes/ .

Andrychowicz, Amy. "Seed Starting 101: The Ultimate Guide to Growing Plants from Seed." *Get Busy Gardening*, 16 Mar. 2017, https://getbusygardening.com/growing-seeds/

Angelo. "What Is Companion Planting and How Does It Work?" *Deep Green Permaculture*, 17 Aug. 2009, https://deepgreenpermaculture.com/2009/08/17/companion-planting/#:~:text=Companion%20planting%20is%20the%20practice.

"Choosing the Right Location for Your Vegetable Garden." *Newsroom*, 7 Apr. 2020, https://sebsnjaesnews.rutgers.edu/2020/04/choosing-the-right-location-for-your-vegetable-garden/

"Companion Planting Chart for Vegetable Garden: Tomatoes, Potatoes, and More! | Guide to Companion Planting | the Old Farmer's Almanac." *Www.almanac.com*, www.almanac.com/companion-planting-guide-vegetables#:~:text=The%20Companion%20Planting%20Chart%20lists.

Hailey, Logan. "15 Companion Planting Mistakes to Avoid This Season." *All about Gardening*, 22 June 2022, https://www.epicgardening.com/companion-planting-mistakes/

"History of Companion Planting – How Did Companion Planting Start." *Gardening Know How*, 22 Mar. 2022, https://blog.gardeningknowhow.com/tbt/history-of-companion-planting/

"How to Mix Organic Fertilizer with Soil – Foliar Garden." *Foliargarden.com*, https://foliargarden.com/how-to-mix-organic-fertilizer-with-soil/

"How to Use Cover Crops to Improve Soil." *FineGardening*, 23 Sept. 2020, www.finegardening.com/project-guides/gardening-basics/how-to-use-cover-crops-to-improve-soil.

https://www.facebook.com/marthastewart. "The Difference between Deadheading and Pruning – and How to Use Each for Healthier Plants and Flowers." *Martha Stewart*, www.marthastewart.com/8041967/deadheading-pruning-differences.

https://www.facebook.com/thespruceofficial. "Companion Plants Repel Garden Pests and Attract Beneficial Insects." *The Spruce*, 2019, www.thespruce.com/companion-planting-1402735.

https://www.facebook.com/WebMD. "Benefits of Companion Planting." *WebMD*, www.webmd.com/a-to-z-guides/benefits-of-companion-planting#:~:text=One%20of%20the%20few%20companion.

https://www.howstuffworks.com/hsw-contact.htm. "HowStuffWorks Answers Your Gardening Questions." *HowStuffWorks*, 21 Aug. 2007, home.howstuffworks.com/gardening/garden-design/gardening-questions-answered.htm.

Judd, Angela. "Garden Troubleshooting Guide: How to Identify & Solve Common Garden Problems." *Growing in the Garden*, 7 Jan. 2022, https://growinginthegarden.com/garden-troubleshooting-guide-how-to-identify-solve-common-garden-problems/

margo. "The Complete 2023 Organic Fertilizer's Guide for Plants." *HomeBiogas*, 18 Jan. 2023,

"Organic Fertilizer vs. Chemical Fertilizer | Kellogg Garden Organics™." *Kellogggarden.com*, https://kellogggarden.com/blog/fertilizer/the-advantages-of-organic-fertilizers-over-chemical-fertilizers/

Poindexter, Jennifer. "10 Tips to Harvest Your Garden Vegetables Perfectly and on Time." *MorningChores*, 4 Apr. 2018, https://morningchores.com/harvesting-your-garden/

"Should You Plant Seeds or Plants in Your Garden? • Gardenary." *Gardenary*, www.gardenary.com/blog/should-you-plant-seeds-or-plants-in-your-garden.

"Soil Preparation: How Do You Prepare Garden Soil for Planting?" *Almanac.com*, www.almanac.com/soil-preparation-how-do-you-prepare-garden-soil-planting.

Walliser, Jessica. "Plant Covers to Protect the Garden from Pests and Weather." *Savvy Gardening*, 29 Apr. 2022, https://savvygardening.com/plant-covers/

When and How to Water Your Seedlings and Seeds the Right Way. 13 Feb. 2023, www.gardeningchores.com/watering-seedlings/.

(N.d.). Businesscasestudies.co.uk. https://businesscasestudies.co.uk/amazing-history-of-succulents-where-do-they-come-from/

(N.d.). Gardenia.net. https://www.gardenia.net/guide/great-shrubs-as-companion-plants-for-your-succulents

(N.d.). Weekand.com. https://www.weekand.com/home-garden/article/pollinate-cactus-18060952.php

(N.d.-a). Masterclass.com. https://www.masterclass.com/articles/bunny-ear-cactus-guide#:~:text=The%20bunny%20ear%20cactus%20is,or%20in%20outdoor%20rock%20gardens.

(N.d.-b). Masterclass.com. https://www.masterclass.com/articles/bunny-ear-cactus-guide#:~:text=In%20order%20to%20keep%20your,moisture%20and%20drain%20excess%20water.

(N.d.-c). Gardenia.net. https://www.gardenia.net/plant/epiphyllum-oxypetalum#:~:text=Water%20regularly%20during%20the%20growing,damage%20or%20kill%20the%20plant.

(N.d.-d). A-z-animals.com. https://a-z-animals.com/plants/old-lady-cactus/#:~:text=This%20plant%20thrives%20outside%20in,inside%20during%20winter%20and%20fall

14 fragrant succulents & cacti (you will love). (2022, February 19). AskGardening. https://askgardening.com/fragrant-succulents-cactus/

25 beautiful cactus garden ideas. (n.d.). Trees.com. https://www.trees.com/gardening-and-landscaping/cactus-garden-ideas

29 types of cactus (with pictures and names) – identification guide. (2021, October 11). Leafy Place. https://leafyplace.com/types-of-cacti/

Abramson, A., & Milbrand, L. (2022, April 14). How to propagate succulents from leaves or stems. Real Simple. https://www.realsimple.com/home-organizing/gardening/indoor/how-to-propagate-succulents

Acanthocalycium glaucum. (n.d.). https://www.cactus-art.biz/schede/ACANTHOCALYCIUM/Acanthocalycium_glaucum/Acanthocalycium_glaucum/Acanthocalycium_glaucum.htm

Acanthocereus tetragonus cv. Fairytale castle. (n.d.). http://www.llifle.com/Encyclopedia/CACTI/Family/Cactaceae/6981/Acanthocereus_tetragonus_cv._Fairytale_castle

Apples and pears: espalier pruning and training / RHS Gardening. (n.d.). Org.uk. https://www.rhs.org.uk/fruit/apples/training-espalier

Baldwin, D. L. (2020, February 12). How succulents combat global warming. Debra Lee Baldwin. https://debraleebaldwin.com/succulent-plants/succulents-combat-global-warming/

Baldwin, D. L. (2021, March 11). What makes succulents SUCCULENT? Debra Lee Baldwin. https://debraleebaldwin.com/cactus/science-of-succulence/

Balogh, A. (2017, May 17). How to plant succulents + 8 growing tips – garden design. Gardendesign.com; Garden Design Magazine. https://www.gardendesign.com/succulents/planting.html

Beaulieu, D. (2013, January 20). What's the difference between Cacti and succulents? The Spruce. https://www.thespruce.com/difference-between-cacti-and-succulents-3976741

Blumberg, P. O. (2020, June 16). 5 expert tips to take care of your cacti. Southern Living. https://www.southernliving.com/garden/plants/cactus-care-tips

Blumberg, P. O. (2020, June 16). 5 expert tips to take care of your cacti. Southern Living. https://www.southernliving.com/garden/plants/cactus-care-tips

Boeckmann, C. (n.d.). Aloe Vera. Almanac.com. https://www.almanac.com/plant/aloe-vera

Box, S. (n.d.). Common Pests & Diseases Treatment for Succulents. Succulents Box. https://succulentsbox.com/pages/common-pests-diseases

Box, S. (n.d.). How to care for your cactus and succulent arrangements. Succulents Box. https://succulentsbox.com/blogs/blog/how-to-plant-cacti-and-succulents-together

Brazilian prickly pear (Brasiliopuntia brasiliensis) Flower, Leaf, Care, Uses. (n.d.). PictureThis. https://www.picturethisai.com/wiki/Brasiliopuntia_brasiliensis.html

Brazilian prickly pear Care (Watering, Fertilize, Pruning, Propagation). (n.d.). PictureThis. https://www.picturethisai.com/care/Brasiliopuntia_brasiliensis.html

Brown, D. L. (n.d.). Cacti and succulents. Umn.edu. https://extension.umn.edu/houseplants/cacti-and-succulents

Brown, D. L. (n.d.). Cacti and succulents. Umn.edu. https://extension.umn.edu/houseplants/cacti-and-succulents

Brown, D. L. (n.d.). Cacti and succulents. Umn.edu. https://extension.umn.edu/houseplants/cacti-and-succulents

Bunny Ears Cactus. (n.d.). Planterina. https://planterina.com/blogs/indoor-plant-care/bunny-ears-cactus

Burro's tail, Sedum morganianum. (n.d.). Wisconsin Horticulture. https://hort.extension.wisc.edu/articles/burros-tail-sedum-morganianum/

Cacti and succulent problems. (2019, March 1). About The Garden Magazine. https://www.aboutthegarden.com.au/cacti-and-succulent-problems/

Cactus care & tips. (2022, March 13). Smith Rock Cactus Company. https://smithrockcactuscompany.com/cactus-care-tips/

Cactus care & tips. (2022, March 13). Smith Rock Cactus Company. https://smithrockcactuscompany.com/cactus-care-tips/

Cactus care. (2007, August 15). HowStuffWorks. https://home.howstuffworks.com/cactus-care1.htm

Cactus care: How to water cacti (cactuses)? (2018, August 31). Terrarium Planting Guide. https://plantinterrarium.com/cactus-care-how-to-water-cacti-cactuses/

Cactus corner project – history. (n.d.). Theinkrag.com. https://www.theinkrag.com/cactus_corner_project/historycc.html

Cactus. (n.d.). Mcgill.Ca

Carberry, A. (2005, January 29). How to grow a Cactus. WikiHow. https://www.wikihow.com/Grow-a-Cactus

Caring for Cacti & succulents – indoor plants – Westland. (2018, February 5). Garden Health. https://www.gardenhealth.com/advice/plants-flowers/how-to-care-for-cacti-and-succulents

Caring for Cacti & succulents – indoor plants – Westland. (2018, February 5). Garden Health. https://www.gardenhealth.com/advice/plants-flowers/how-to-care-for-cacti-and-succulents

Carnegiea gigantea | Landscape Plants | Oregon State University. (n.d.). https://landscapeplants.oregonstate.edu/plants/carnegiea-gigantea

Common Succulent diseases and pests to look out for. (n.d.). Plants For All Seasons. https://www.plantsforallseasons.co.uk/blogs/succulent-care/common-succulent-diseases-and-pests-to-look-out-for

Cooperative extension: Garden and yard. (2011, December 2). Cooperative Extension: Garden and Yard. https://extension.umaine.edu/gardening/manual/propagation/plant-propagation/

Cowen, J. (2021, May 2). 5 best succulent and cactus fertilizer + how to fertilize. The Yard and Garden. https://theyardandgarden.com/best-succulent-and-cactus-fertilizer/

Cox, M. (n.d.). How to prune topiary. Saga.co.uk. https://www.saga.co.uk/magazine/home-garden/gardening/advice-tips/pruning/how-to-prune-topiary

Creative Wilder. (2021, February 26). 4 benefits of propagating plants. Pop Wilder Website. https://www.popwilder.com/post/4-benefits-of-propagating-plants

D, J. (2023). 132 Different Types of Cacti listed in A to Z Photo Database. Home Stratosphere.

Edinburg Scenic Wetlands, & World Birding Center. (n.d.). Cochineal insect: The natural dye. Edinburg Scenic Wetlands and World Birding Center.

Eg. (2021). Trichocereus thelegonus / Echinopsis thelegona Friedrich & Rowley. Trichocereus.net – Cactus Seeds and Books! https://trichocereus.net/trichocereus-thelegonus-echinopsis-thelegona-friedrich-rowley/

Epiphyllum oxypetalum. (n.d.). http://www.llifle.com/Encyclopedia/CACTI/Family/Cactaceae/8223/Epiphyllum_oxypetalum

Espiritu, K. (2019, November 6). How to grow a ponytail palm outdoors. Epic Gardening. https://www.epicgardening.com/growing-ponytail-palm-outdoors/

Forbes, N. (2022, July 15). Growing cacti & succulents indoors. Dennis' 7 Dees | Landscaping Services & Garden Centers; Dennis' 7 Dees Landscaping & Garden Centers. https://dennis7dees.com/growing-cacti-succulents-indoors/

Foster, N. (2022, May 21). Indoor cactus care for beginners (2023 guide). Joy Us Garden | Care, Propagation, and Pruning; Joy Us Garden. https://www.joyusgarden.com/indoor-cactus-care/

Full guide on cactus diseases, pests and treatments. (2018, October 26). Terrarium Planting Guide. https://plantinterrarium.com/full-guide-on-cactus-diseases-pests-and-treatments/

García, A. (2022, October 24). The importance of Nopal cactus to Mexicans. Mansion Mauresque. https://mansionmauresque.com/the-importance-of-nopal-cactus-to-mexicans/

Gaumond, A. (2023, July 31). The cactus: A guide to its meaning, symbolism & cultural significance. Petal Republic. https://www.petalrepublic.com/cactus-meaning/

Gilmer, M. (2015, January 16). The Desert Sun. The Desert Sun. https://www.desertsun.com/story/life/home-garden/maureen-gilmer/2015/01/16/san-pedro-cactus-ornamental-maureen-gilmer/21789499/

Gotter, A. (2017, May 25). Nopal cactus: Benefits, uses, and more. Healthline. https://www.healthline.com/health/nopal

Grant, B. L. (2012, December 30). Starfish flower cactus: Tips for growing starfish flowers indoors. Gardening Know How. https://www.gardeningknowhow.com/ornamental/cacti-succulents/starfish-flower/growing-starfish-flowers.htm

Grant, B. L. (2014, September 12). Established plants are tall and leggy: What to do for leggy plant growth. Gardening Know How. https://www.gardeningknowhow.com/plant-problems/environmental/established-plants-leggy.htm

Grant, B. L. (2018, May 19). Walking stick cholla info: Tips on caring for walking stick chollas. Gardening Know How. https://www.gardeningknowhow.com/ornamental/cacti-succulents/cholla-cactus/caring-for-walking-stick-chollas.htm

Gregarious Inc. (n.d.-a). Old Man Cactus. Greg.App. https://greg.app/plant-care/old-man-cactus-cephalocereus-senilis

Gregarious Inc. (n.d.-b). What Soil is Best for a Christmas Cactus? Greg.App. https://greg.app/blog/cactus/what-soil-is-best-for-a-christmas-cactus/

Growing aloe. (n.d.). Miraclegro.com. https://miraclegro.com/en-us/indoor-gardening/growing-aloe.html

Gymnocalycium saglionis subs. tilcarense. (n.d.). http://www.llifle.com/Encyclopedia/CACTI/Family/Cactaceae/12119/Gymnocalycium_saglionis_subs._tilcarense

Hailey, L. (2022, June 22). 15 companion planting mistakes to avoid this season. All About Gardening.

Hassani, N. (2023, February 28). 31 types of succulents worth growing. The Spruce. https://www.thespruce.com/types-of-succulents-7090763

Hicks-Hamblin, K. (2021, September 29). The scientifically-backed benefits of companion planting. Gardener's Path. https://gardenerspath.com/how-to/organic/benefits-companion-planting/

Homegrown Garden. (n.d.). Which succulents can be planted together? Homegrown Garden. https://homegrown-garden.com/blogs/blog/which-succulents-can-be-planted-together

How to care for a Christmas Cactus: A complete guide. (2021, November 12). The Sill. https://www.thesill.com/blog/christmas-cactus

How to care for a Pencil Cactus plant. (n.d.). Greeneryunlimited.Co. https://greeneryunlimited.co/pages/pencil-cactus-care

How to grow a Christmas cactus. (n.d.). Miraclegro.com. https://miraclegro.com/en-us/indoor-gardening/how-to-grow-a-christmas-cactus.html

How to grow cacti & succulents. (n.d.). Horticulture Magazine. https://horticulture.co.uk/succulents/

How, G. K. (2016, April 9). Fertilizing cactus plants: When and how to fertilize A cactus. Gardening Know How. https://www.gardeningknowhow.com/ornamental/cacti-succulents/scgen/fertilizing-cactus-plants.htm

How, G. K. (2016, April 9). Fertilizing cactus plants: When and how to fertilize A cactus. Gardening Know How. https://www.gardeningknowhow.com/ornamental/cacti-succulents/scgen/fertilizing-cactus-plants.htm

Hughes, M. (2021, September 3). How to fix leggy plants for a lush indoor garden. Better Homes & Gardens. https://www.bhg.com/gardening/houseplants/care/fix-leggy-houseplants/

Iannotti, M. (2018, October 1). Crown of Thorns (Euphorbia milii): Care & Grow Guide. The Spruce. https://www.thespruce.com/crown-of-thorns-plant-4175182

Insect pests of cacti and succulents grown as house plants. (n.d.). Missouribotanicalgarden.org. https://www.missouribotanicalgarden.org/gardens-gardening/your-garden/help-for-the-home-gardener/advice-tips-resources/pests-and-problems/insects/mealybugs/insect-pests-of-cacti-and-succulents

Jade plant: How to grow and care for Jade plants – garden design. (2020, September 22). Gardendesign.com; Garden Design Magazine. https://www.gardendesign.com/succulents/jade-plant.html

Jagdish. (2021, May 10). Pruning in agriculture – benefits, tips, and ideas. Agri Farming. https://www.agrifarming.in/pruning-in-agriculture-benefits-tips-and-ideas

Jamie. (2021, September 3). How to fix leggy plants and seedlings + prevention tips. WhyFarmIt. https://whyfarmit.com/how-to-fix-leggy-plants/

Janie. (2020, October 5). Fertilizer for cactus: When, how and in what ratio. Succulent Alley. https://succulentalley.com/fertilizer-for-cactus/

Janie. (2020, September 29). How to plant cactus in ground: A step-by-step guide. Succulent Alley. https://succulentalley.com/how-to-plant-cactus-in-ground/

Janie. (2023, January 31). Can cactus and succulents be planted together? Succulent Alley. https://succulentalley.com/can-cactus-and-succulents-be-planted-together/

Johnson's Beehive Cactus (Echinomastus johnsonii). (n.d.). iNaturalist United Kingdom. https://uk.inaturalist.org/taxa/871500-Echinomastus-johnsonii

Lalko, A. (2023). Starfish Cactus: Plant Guide. Independently Published.

Limiterd, G. L. (n.d.). Ball cactus (Parodia magnifica) Flower, Leaf, Care, Uses – PictureThis. PictureThis. https://www.picturethisai.com/wiki/Parodia_magnifica.html

Mammillaria dixanthocentron. (n.d.). http://www.llifle.com/Encyclopedia/CACTI/Family/Cactaceae/19098/Mammillaria_dixanthocentron

McCarthy, K. (2018, July 3). Dividing succulents – propagating succulents. The Succulent Eclectic. https://thesucculenteclectic.com/dividing-succulents-propagating/

McKie, C. (2022). The Delicate Beauty of the Lady Finger Cactus. Houseplant Central. https://houseplantcentral.com/lady-finger-cactus/

McKie, C. (2022, September 20). The Starfish cactus: A unique beauty. Houseplant Central. https://houseplantcentral.com/starfish-cactus/

Mexican snow ball Care (Watering, Fertilize, Pruning, Propagation). (n.d.). PictureThis. https://www.picturethisai.com/care/Echeveria_elegans.html

Mexican Snowball plant care & growing basics: Water, light, soil, propagation etc. (n.d.). Myplantin.com. https://myplantin.com/plant/392

Miller, R. (2019, August 19). Where do most succulents come from? Succulent City; SucculentCity.com. https://succulentcity.com/where-do-most-succulents-come-from/

Miller, R. (2022, August 1). How to trim/ prune succulents successfully (an easy guide from A succulent expert). Succulent City; SucculentCity.com. https://succulentcity.com/how-to-trim-succulents/

Miller, R. (2022, August 7). 5 organic succulent fertilizers for naturally feeding your garden. Succulent City; SucculentCity.com. https://succulentcity.com/organic-succulent-fertilizer/

Miller, R. (2023). The Old Lady Cactus 'Mammillaria Hahniana' Succulent City. https://succulentcity.com/old-lady-cactus/

Old lady cactus Care (Watering, Fertilize, Pruning, Propagation). (n.d.). PictureThis. https://www.picturethisai.com/care/Mammillaria_hahniana.html

Old Lady cactus Plant Care & Growing Basics: Water, Light, Soil, Propagation, etc. (n.d.). Myplantin.com. https://myplantin.com/plant/393

Old man cactus Care (Watering, Fertilize, Pruning, Propagation). (n.d.). PictureThis. https://www.picturethisai.com/care/Cephalocereus_senilis.html

Old man cactus growing and plant care guide – good life vibez. (n.d.). Goodlifevibez.com. https://goodlifevibez.com/old-man-cactus-growing-and-plant-care-guide/

Old man cactus. (2016, January 5). Horticulture Unlimited. https://horticultureunlimited.com/plant-guide/old-man-cactus/

Opuntia microdasys bunny ears, angel's-wings PFAF plant database. (n.d.). Pfaf.org. https://pfaf.org/User/Plant.aspx?LatinName=Opuntia+microdasys

Oreocereus trollii old man of the mountain. (n.d.). Planet Desert. https://planetdesert.com/products/oreocereus-trollii-old-man-of-the-mountain-cactus-cacti-nice-real-live-plant

OREOCEREUS TROLLII. Old Man Of The Mountain Cactus. (n.d.). Thepalmtreecompany. https://www.thepalmtreecompany.com/product-page/oreocereus-trollii-old-man-of-the-mountain-cactus-1

Paschall, T. P. (2022, March 22). 5 ways to makeover an overgrown garden. Best Pick Reports. https://www.bestpickreports.com/blog/post/5-ways-to-makeover-an-overgrown-garden/

Pino, M. (2023, June 13). How to plant, grow, and care for queen of the Night Flower. Planet Natural. https://www.planetnatural.com/queen-of-the-night-flower/

Place, V. (2012). Page Not Found. Lulu.com. https://www.hgtv.com/outdoors/flowers-and-plants/how-to-plant-a-cactus-container-garden

Plant ID: Flowers and Foliage: Zebra Plant – Florida Master Gardener Volunteer Program – University of Florida, Institute of Food and Agricultural Sciences. (n.d.). https://gardeningsolutions.ifas.ufl.edu/mastergardener/outreach/plant_id/flowers_indoor/zebra_plant.html

PlantIn. (n.d.). Notocactus Magnificus Plant Care & Growing Basics: Water, Light, Soil, Propagation etc. | PlantIn. https://myplantin.com/plant/6212

Rahman, A. (2021, April 12). Signs of under-watered cactus (and how to revive it). Garden For Indoor.

Rana, A. (2023, February 10). Cactus care 101: Everything you need to know about having a cactus. Planet Desert. https://planetdesert.com/blogs/news/how-to-care-for-a-cactus

risk of overwatering of cacti and succulents – Google Search. (n.d.). Google.com. https://www.google.com/search?q=risk+of+overwatering+of+cacti+and+succulents&oq=risk+of+overwatering+of+cacti+and+succulents&aqs=chrome..69i57.27427j0j7&sourceid=chrome&i.e.,=UTF-8

risk of underwatering of cacti and succulent plants – Google Search. (n.d.). Google.com. https://www.google.com/search?q=risk+of+underwatering+of+cacti+and+succulent+plants&oq=risk+of+underwatering+of+cacti+and+succulent+plants&aqs=chrome..69i57j33i16012.18649j0j7&sourceid=chrome&i.e.,=UTF-8

San Pedro column cactus Care (Watering, Fertilize, Pruning, Propagation). (n.d.-a). PictureThis. https://www.picturethisai.com/care/Echinopsis_pachanoi.html

San Pedro column cactus Care (Watering, Fertilize, Pruning, Propagation). (n.d.-b). PictureThis. https://www.picturethisai.com/care/Echinopsis_pachanoi.html

Sánchez, M. (2017, November 9). What are the parts of the cactus, and what functions do they have? Jardineria On. https://www.jardineriaon.com/en/partes-del-cactus-y-sus-funciones.html

Schiller, N. (2019, February 3). Propagating succulents in 5 easy steps. Gardener's Path. https://gardenerspath.com/how-to/propagation/succulents-five-easy-steps/

Schiller, N. (2022, July 7). How to grow succulents outdoors in the garden. Gardener's Path. https://gardenerspath.com/plants/succulents/grow-garden-succulents/

Sears, C. (2020, June 17). How to grow and care for chocolate soldier plant. The Spruce. https://www.thespruce.com/chocolate-soldier-plant-profile-5024790

Sears, C. (2021a, April 13). How to grow and care for Mexican snowballs. The Spruce. https://www.thespruce.com/mexican-snowballs-echeveria-elegans-profile-5120250

Sears, C. (2021b, June 30). How to grow and care for the bunny ear cactus. The Spruce. https://www.thespruce.com/bunny-ear-cactus-guide-5190802

Secuianu, M. (2020a, November 3). Old Lady Cactus guide: How to grow & care for "Mammillaria Hahniana." GardenBeast. https://gardenbeast.com/old-lady-cactus-guide/

Secuianu, M. (2020b, November 18). Brasiliopuntia Brasiliensis Guide: How to grow & care for "Brazilian Prickly Pear." GardenBeast. https://gardenbeast.com/brasiliopuntia-brasiliensis-guide/

Sher, S. (2021, September 30). 25 types of succulents that make great houseplants. Bob Vila; BobVila.com. https://www.bobvila.com/articles/types-of-succulents/

Spengler, T. (2018, February 24). Managing large shrubs – learn how to trim an overgrown shrub. Gardening Know How. https://www.gardeningknowhow.com/ornamental/shrubs/shgen/trimming-overgrown-shrub.htm

Staff, G. (2019, May 6). Austrocephalocereus dybowskii – Giromagi Cactus and Succulents. Giromagi Cactus and Succulents. https://www.giromagicactusandsucculents.com/austrocephalocereus-dybowskii-giromagi-cactus-succulents/

StaffSimone. (2022, May 30). Acanthocalycium thionanthum – Giromagi Cactus and Succulents. Giromagi Cactus and Succulents. https://www.giromagicactusandsucculents.com/acanthocalycium-thionanthum/

Stamp, E., & McLaughlin, K. (2018, September 12). How to care for succulents (and not kill them): 9 plant care tips. Architectural Digest. https://www.architecturaldigest.com/story/how-to-care-for-succulents

Stephens, T. (2019, May 29). A guide to watering cactus. Cactus Culture – Australia's Premium Online Cactus Shop. https://cactusculture.com.au/learning-centre/cactus-watering-guide

Stephens, T. (2022, August 29). Does a cactus need water? Cactus Culture – Australia's Premium Online Cactus Shop. https://cactusculture.com.au/learning-centre/does-a-cactus-need-water

Strawberry cactus care (watering, fertilizing, pruning, propagation). (n.d.-a). PictureThis. https://www.picturethisai.com/care/Mammillaria_dioica.html

Strawberry cactus care (watering, fertilizing, pruning, propagation). (n.d.-b). PictureThis. https://www.picturethisai.com/care/Mammillaria_dioica.html

Succulents Australia. (n.d.). Thimble Cactus – Mammillaria gracilis fragilis. https://www.succulents-australia-sales.com/products/thimble-cactus?variant=31372756222083

Taylor, L. H. (2016, June 30). 21 best cactus plants to grow in your garden. The Spruce. https://www.thespruce.com/best-cactus-to-plant-in-garden-4059807

Temperature and humidity. (2021, December 14). Plnts.com. https://plnts.com/en/care/doctor/temperature-and-humidity

The complete guide to trimming succulents. (n.d.). Lula's Garden. https://www.lulasgarden.com/blogs/all-blogs/the-complete-guide-to-trimming-succulents

The Ruth Bancroft Garden & Nursery. (2023, March 3). Echinopsis formosa – The Ruth Bancroft Garden & Nursery. https://www.ruthbancroftgarden.org/plants/echinopsis-formosa/

Tuttle, C. (2021, May 6). How to care for succulents indoors. Succulents and Sunshine. https://www.succulentsandsunshine.com/guide-growing-succulents-indoor-house-plants/

Ultimate guide: How to fight cactus diseases and pests. (2021, November 7). CactusWay. https://cactusway.com/ultimate-guide-how-to-fight-cactus-diseases-and-pests/

Vanderlinden, C. (2011, January 18). How to prevent leggy vegetable seedlings. The Spruce. https://www.thespruce.com/preventing-your-seedlings-from-getting-leggy-2539979

Vanderzeil, G. (2018, May 15). What's wrong with my succulent?! Collective Gen. https://collectivegen.com/2018/05/whats-wrong-succulent/

VanZile, J. (2008, October 23). How to grow and care for Kalanchoe. The Spruce. https://www.thespruce.com/growing-kalanchoe-plants-1902982

VanZile, J. (2008, September 11). How to grow and care for indoor cactus. The Spruce. https://www.thespruce.com/how-to-grow-cactus-1902954

VanZile, J. (2008, September 11). How to grow and care for indoor cactus. The Spruce. https://www.thespruce.com/how-to-grow-cactus-1902954

VanZile, J. (2009a, May 19). Ponytail Palm Plant Profile. The Spruce. https://www.thespruce.com/grow-beaucarnea-recurvata-1902886

VanZile, J. (2009b, November 17). How to care for Jade plants: Indoor growing guide. The Spruce. https://www.thespruce.com/grow-jade-plants-indoors-1902981

VanZile, J. (2022a). How to Grow and Care for Rat Tail Cactus. The Spruce. https://www.thespruce.com/aporocactus-flagelliformis-definition-1902538

VanZile, J. (2022b). How to Grow and Care for Ball Cactus. The Spruce. https://www.thespruce.com/grow-parodia-cacti-indoors-1902591

Waddington, E. (2023, May 17). 7 common cacti problems & solutions. Horticulture Magazine. https://horticulture.co.uk/cacti-problems/

west-coast-gardens. (2023, January 18). 5 care tips to keep your cactus happy. West Coast Gardens. https://www.westcoastgardens.ca/blogs/tips-inspiration/5-care-tips-to-keep-your-cactus-happy

Wet, & Forget. (2019, February 12). Fixes for the most common succulent pests and diseases. Life's Dirty. Clean Easy. https://askwetandforget.com/fixes-common-succulent-pests-diseases/

Weymouth, M. (2019, January 28). How to use rooting hormone when propagating plants. Martha Stewart. https://www.marthastewart.com/1535873/how-to-use-rooting-hormone

What is integrated pest management (IPM)? (n.d.). Ucanr.edu. https://ipm.ucanr.edu/what-is-ipm/

What is Pruning? Importance, Benefits & Methods of Pruning. (n.d.). Davey.com. https://blog.davey.com/what-is-pruning-the-importance-benefits-and-methods-of-pruning/

What is the best fertilizer for your cactus? (2021, February 23). CactusWay. https://cactusway.com/what-is-the-best-fertilizer-for-your-cactus/

White, J. (2022, June 23). 57 types of succulents with names and pictures. All About Gardening.

Why is Companion Planting So Important? (2023, June 19). Triangle Gardener Magazine; Triangle Gardener LLC. https://www.trianglegardener.com/why-is-companion-planting-so-important/

Wiley, D. (2015, June 9). How to grow cactus plants in cold-winter climates. Better Homes & Gardens. https://www.bhg.com/gardening/flowers/perennials/growing-cactus-plants-in-cold-climates/

Wolfe, D. (2020, June 3). The best pots for succulents of 2023. Bob Vila; BobVila.com. https://www.bobvila.com/articles/best-pots-for-succulents/

WoS. (2022). Epiphyllum hookeri (Hooker's Orchid Cactus). World of Succulents. https://worldofsucculents.com/epiphyllum-hookeri/

Yang, K. (2021, November 4). 5 surprising facts about cactus sustainability. Pricklee Cactus Water. https://pricklee.com/blogs/learn/cactus-sustainability

Young, C. (2023, May 8). The best tips on growing a San Pedro cactus. The Planted Pot. https://theplantedpot.co.nz/blogs/plant-care/san-pedro-cactus

www.ingramcontent.com/pod-product-compliance
Lightning Source LLC
Chambersburg PA
CBHW051855160426
43209CB00006B/1310